# Sweet Dreamer

Lu Xiao Chang

# CONTENTS

# CHAPTER 1

"Not necessary."

The shot rang out across the gas station as droplets of blood splattered across my face. The arms of the murderer unwound themselves from my body as he fled, leaving me with a gaping hole where my heart used to be. I stood there, staring at my heart as it took its final, agonizing breaths with tears running down its face.

"Hey, Haz!" Anastasia said. Her voice tore me out of my trance. My eyes met the taupe inner walls of my locker, my fingers still locked around my Macbeth book. The cover felt damp under my fingertips. "Hey... you good?"

I gave her a mechanical nod and then glanced over in her direction. She laughed with no remorse. "You look beyond tired."

The hell hole I called school was the reason for the lack of energy coursing through my body—my eyelids were like lead, and my head throbbed in a painful ache. I stayed up all night studying for my Spanish test—why did I even have to take Spanish when I couldn't speak?

The bell rang, and she raised her eyebrows at my lack of acknowledgment. I was busy pushing that dark memory of mine to the back of my head. "Hello? You all right? We gotta go, or we're going to be late."

I quickly grabbed the various items I would need for my classes before lunch and followed her through the glossy beige and white tiled halls. Could I nap today at all? Psychology was probably my best bet since that class was a bore—the teacher didn't seem to care about what he was teaching. So I wouldn't miss much except maybe a PowerPoint that looked like a third-grader prepared it.

I entered Homeroom and sank into my seat. Anastasia sat next to Blake, who glanced at me. "You don't look so good. What's up?" I shrugged.

He turned his attention to Ana. "What's up with her?" Anastasia and Blake were my closest friends out of everybody in my friend group. We grew up on the same block, went to the same schools since kindergarten, and had the same paren—nope.

Jeez, my brain was fried.

"She's just exhausted, I think," Anastasia said.

He frowned. "Want me to run and get you a coffee or something?" I shook my head. He didn't care that he'd get in trouble for leaving and was so self-sacrificing when it came to Ana and me; he'd risk his skin to make sure we were comfortable. It was unnaturally noble, and most boys his age weren't like that—at least not in our school.

"How're you gonna pass the Spanish test today if you're this tired?" he pressed.

I signed, "I'll take a nap later—during lunch." And, Psychology... I knew he would disapprove if I told them that part.

His face was the embodiment of concern. "So you're not gonna eat?" Dear Lord, please leave me alone. This was becoming 20 questions when all I wanted to do was close my eyes to soothe my friggin headache.

I sighed and signed, "I'll eat quickly and take a nap."

He shook his head. "Next time you pull a stunt like this and don't get your proper sleep time, I'll have to beat you. Do you understand?" I laughed and rolled my eyes.

Thanks to me, Anastasia and Blake both learned to understand American Sign Language. As a child, I stopped speaking, so I could only communicate through ASL (before I could write fluently). And since your brain is a sponge at that age, they got with the program quickly. Of course, they didn't sign back to me since they could still verbally communicate, but they understood me.

And it's all because of one bullet that I could no longer speak.

The bell rang, causing chairs to screech as they were pushed back and people to grumble. Finally, the first period—the great evil—was upon us. Slightly slumped, I walked out of class and down the hall to the worst first-period class ever to be invented: Physics. I could only handle so much torture in the morning.

My school had its cons, like the students and first-period physics, but it wasn't too terrible. Some of the gems in my school were the teachers. They were understanding of my situation and didn't count class participation against me. If I wanted to answer a question, I'd write it down on paper, raise my hand, and then hold up the answer. But other than small instances like that, I tended to keep to myself in class. And it was acceptable.

One of the cons was my peers, who were stereotypical regarding clique formation. The popular people always tended to be white with blue eyes and blonde hair and were most likely to win prom queen and king. Typical. A lot of them were rich, or at least extremely wealthy. And they partied—a lot.

But although the students here went by the usual class separation in school, the bullying here was different—or maybe just new. There was fake-ness and talking behind your back rather than anything physical. So even if people acted like they thought it was cool I could sign and then talked trash about me, I wouldn't know because no one was brave enough to say anything to anyone's face. I didn't mind that, in all honesty. It helped me get through school without feeling ashamed for being unable to speak.

"... the terminal velocity can be determined with this equation..."

I pulled my head out of autopilot, trying my hardest to focus on the board. Physics dragged on, and my eyes were growing heavier by the minute. I bounced my leg, but that seemed to make me sleepier. Some moments I could see the board, and for others, I was in darkness. Every time I was, I shook myself awake.

"Hazel, are you all right?" my teacher whispered. My eyes snapped open, and I looked up at her, nodding.

"Why are you so tired, hun?" she asked. I began to sign and then stopped to write in my notebook. I stayed up late studying for my Spanish test. I presented the note to her.

My teacher nodded sympathetically and said, "I understand. Try to stay awake, though, okay? I don't want you to miss any of this." I nodded and mouthed, "Sorry." She nodded and rubbed my back, returning to the front of the class. I realized some eyes were on me even though a problem was left on the board.

"Class, does anybody have an answer?" Ms. Iaradi asked. People returned their focus to her, and I relieved a small sigh.

It was going to be a long day.

***

By the time lunch rolled around, I felt slightly better. I got my nap in during Psychology without anyone even noticing. The only problem was my brain decided it needed some more sleep still. So when I plopped into my seat at lunch, I scarfed down my food.

"Slow down," Mindy laughed. I didn't even bother to sign my response. Mindy was one of my other friends and the only other girl in our group. I loved her to death, and she was practically as close to me as I was to Ana—but at that moment, I wanted to choke her. I needed to get my second nap in before burning all my bridges.

They talked amongst themselves as I finally finished my food and let my head rest on the table. Then, with my eyes closed, my ears finally tuned into the conversation my friends were having.

"You didn't know this?" Ana asked. "Grey broke up Sasha Bernstein. She was cheating on him with Evan Clarke from sophomore year chem class."

They caught my attention. Sasha Bernstein was in my English class with Mindy.

"Who are these people?" Andy asked. I could tell he was already bored. He was never one for liking drama.

"Just some people from the popular crowd in this school," Ana replied. "The whole school is freaking out about this."

"Why?" Arnold joined in. I could already tell he was nervous; the popular kids were like an allergy to him. But, unlike Andy, he liked to know, so he knew which direction to run when things went down.

"'Cause Grey is cute, dude—we gotta discuss that he's free now," Mindy gushed. I smiled and rolled my eyes.

"You don't think any of us guys are cute?" Andy asked, taking fake offense.

"No," Mindy answered. "You're all too nerdy."

"Ironic coming from the girl apart of this squad!" Blake teased.

"Shut up!" she exclaimed with a wholesome laugh. "Now help me get with Grey."

"Do you actually wanna get with him?" Ana couldn't keep the surprise out of her voice.

"No, of course not!" I peeked up at Mindy, who rolled her eyes exasperated, earning a laugh from the entire table, including myself. "He's throwing a celebratory party tomorrow, though. The whole grade is invited. Even us losers. I wanna get as close as I can to him."

"How is the whole grade invited? Is he rich, living in a big house?" my friend, Arnold, asked.

"Yep," Mindy confirmed. "Filthy stinkin' rich."

"Why's it always the rich that are popular?" Blake asked.

Soon enough, their conversation finally faded away, and my brain got some rest.

***

I couldn't have been happier to pull up to my driveway after school. I shuffled inside my house and gave my mom a quick kiss in the kitchen as Timmy, our golden retriever, jumped around excitedly.

"How was school?" she asked.

"Good," I signed. "I need to take a nap, though."

I ran upstairs, put on my lounging clothes, and threw myself into bed. I earned this nap, being that I aced that test! And my brain was screaming for a way to clear up all the debris there. I slowly drifted off, relishing the feel of my body relaxing into sleep.

"Not necessary."

I jerked awake with an ache in my chest. Taking a deep breath in, I focused on calming my racing heart. This dream and memory were plaguing my mind with a vengeance. I think it's because I've been trying my hardest to avoid being sucked into anyone else's.

I checked the time on my phone (also to see if I had any messages). It felt like I had slept for only four minutes, but it had been two hours. Maybe I could get five more—

"Hazel, dinner's ready!" my mom called from downstairs. I groggily removed the warm covers and headed downstairs into the kitchen. She scooped rice and chili into my bowl as steam rose. The aroma of her spices made my mouth water, and I rushed to the table, shoveling the food into my mouth.

"Jeez, why are you so hungry?" my mom asked with a laugh. She sat down next to me with her serving.

"I didn't eat much lunch today," I signed.

Her eyebrows raised in concern. "Why not?"

I waved her worry away before signing, "I needed to take a nap during lunch because I stayed up too late studying for my test."

She was impassive before she reached for my hand and gave it a light squeeze. "I'm proud of you, baby." I smiled, and she returned it.

I noticed her fidgeting a lot as I continued to eat. Finally, I began to slow down and glanced at her face. Her usual calm demeanor was replaced with nervousness. She didn't make eye contact with me as she began to speak. "Hazel, something good is about to happen in our lives; I know it. I've met someone. Someone who can help me. Someone who will be there for me instead of me selfishly putting my problems on you."

Her words caught me by surprise. Mom had been relying on me a lot after my dad passed away. She'd tell me about things financially and mentally that she never meant to fall on my ears. Of course, I was patient and listened, but I knew my silence was painful for her. The fact that I didn't have a voice to respond to what she said must've gotten to her head. She needed more social interaction than just my hands.

"Someone?" I signed. "Like...?"

"A man I met when I went out with my girlfriends," she said. "We've gotten to know each other more for a while now. I want you to meet him real soon. We're not exactly official yet, but when we do become—"

"When?" Her words burned me more than they should've. "What about Dad?"

"Hazzy..." She let out a heavy sigh as she pinched the bridge of her nose. "Baby, he's gone." She paused, and her eyes watered. "I need to move on."

My eyes met the inside of my bowl to the few morsels I had left and began eating again—slowly—tears threatening to escape. Yeah, Dad was long gone, but I missed him. And yes, she had remained single for years, but how could she just move on?

"Baby, I understand that this is sudden... but I promise you, this is good." I nodded and then shrugged.

"I'm going upstairs now," I signed, picking up my dishes and bringing them to the dishwasher.

"Okay, I'll check up on you later," Mom said.

By the time I made it upstairs, I hardly had time to throw my face down into my pillow before my tears broke loose.

She may be able to move on from Dad, but I will never.

# CHAPTER 2

Darkness shrouded me. An uncomfortable feeling developed in my chest as if I was out of place. Even so, I kept walking. I knew what I had to do.

In the distance, lights glimmered softly. I continued walking towards them until they became brighter, and from an actual source; I was approaching a mansion. Every window glowed yellow-green, which illuminated the exterior as well—literally... the whole place and outdoor garden glowed.

Whose dream am I in now? I wondered. I had more than just the burden of my past haunting me in nightmares. I was also burdened with being sucked into random people's dreams. First, I'd be asleep; I was being dragged into someone else's dreamland the next moment. This ability was with me for as long as I could remember, and my dad used to be the only one who seemed to understand. I didn't tell him

I went into other people's dreams, only that they were so vivid. But he had a knowing look in his eyes. So I knew he got it—that he could empathize even.

I miss him so much.

It took a while to realize I was in other people's dreams. I thought I was just a frequent lucid dreamer—but it turns out that the faces I saw in my dreams, I saw in real life before or after the dream. Of course, when you know a person and then end up dreaming about them, your subconscious could've just put them there. But one day, a writer from out of town visited my elementary school. I already knew what he looked like because we had to read a bit about him before coming to the school. We were also shown his picture, so the fact that I had a dream with him the night prior was no biggie to me.

He started off his speech for us and told us about how most of his stories stemmed from dreams he had that he'd write down every morning when he woke up. Then, much to my surprise and dismay, he described the dream I had about him from his perspective. Everything that happened was the same. We were at a waterfall; I was standing underneath it. I waved at him before jumping into the water below and turning into a mermaid (I was obsessed with the show H2O at that time). He even told them about the beautiful chromatic colors of my scales.

We shared a dream together?

At first, I thought people could share dreams with others. But whenever I recognized people in town after sharing a dream with them, if they saw me, I was rarely acknowledged. It took some time, but it finally clicked in my mind that I was the one who was going into other people's dreams and the only one aware of it.

With that revelation, I started noticing things I hadn't before. Like in my dreams, there was always a sense of belonging and comfort. But when I was in someone else's dreams, I knew I was viewed as a foreign invader. I didn't belong in other people's minds. And I was always desperate to get out of them.

I approached a large golden door and returned my focus to the present. The splendor was magnificent and unreal; a large, golden archway held an equally as large door. The outside of the house seemed to be made of marble and smooth stone. Hesitantly, I knocked on the large double doors. My knocking echoed around the nothingness behind me and inside the house until the door slowly groaned open.

The house was grand. The ceiling arched into a dome high in the sky; the walls consisted of the romantic era style of paintings that brightened up the room. A large staircase ascended and split off to the left and right, with a grand window in the middle where the stairs split. Unfortunately, the window only exposed the darkness beyond.

The banister was lined in gold, and the marble floor was so shiny that it practically glowed.

"Do you have an appointment?" a Butler asked.

"Uh," I said, my own voice slightly scaring me—it always did. "Yes?"

He stared at me blankly for a moment, and I shifted uncomfortably. "Very well then, miss. May I take your jacket?" I looked down at my body. I was wearing a white fur jacket, a skin-tight black sparkling dress that ended right above my knees, sheer stockings, and white high heels.

Whoever this person was, I would have to slap them awake for putting me in something I was so uncomfortable with.

"Yes, please." I turned so he could slide the jacket off my arms. "Right this way, miss." The Butler's shoes clicked as loudly as my heels. He guided me up the stairs, staying on his right at the split. He led me down the red velvet carpeted floors to the end of the hall, where there was a dark door with a light coming from underneath the crack.

"Please proceed, miss," he said. I glanced at him warily and then knocked on the door. I got no response.

"Um... hello?" I asked.

"Come in," a voice called from the other side. I pushed the door open to be met with a room that looked like a study you would most definitely find in every rich, old person's home. The Butler quickly closed it behind me. My eyes roamed around the room. The walls were polished wood with intricate designs carved into them, the windows on the far side of the room were covered in red satin drapes, and a large desk sat in the middle of the windows. A fireplace is a few feet away from the doors.

The chair behind the desk faced the windows, giving the person an overly dramatic air of mystery.

"State your business," the guy said. "Or I'll have to kill you."

The threat fell so hilariously flat that I had to stifle the laughter tickling my chest. The guy's voice was deep but still childlike enough for me to know he wasn't a grown man.

All right, so we're dealing with someone young. Not too bad. I can get out of this, no problem.

"I came to visit you," I said calmly.

He spun around in his chair, his black suit tight and crisp enough to expose how toned he was. His jet-black hair was uniform in his getup. His eyes lit up in surprise; his gaze traveled up and down my body.

"Well, to what do I owe the pleasure?" he asked, standing quickly to approach me. He towered a few inches above me and held his hand out. My fingers met his warm hand as he brought them to his lips for a soft kiss. What movie was he watching before he went to sleep? I feel like I'm in some old black and white film.

"I was..." I thought for a moment. "...in town. Just decided I'd see how you were doing." I honestly had no idea who he was or even what to say.

"Well, my darling, I have many more clients I need to meet with. So I'm going to have to ask you to leave." He sat back down, running his hands through his hair as he grabbed papers off his desk.

"I'm afraid I can't do that," I said. "Not until you wake up."

"Wake up?" he asked, his eyebrows raised.

"Yep." I slammed my hand down on his desk; with a quick motion, I slid all his papers off in a frenzy of black ink and white pages.

"What the hell!" he exclaimed. They floated around, suspended in air—the beauty of the dream world. His chair slid back as he reached up to grab them. The papers slipped through and around his fingers. Then, realizing his efforts were in vain, he emerged through the pages, rage in every crevice of his—dare I admit it—handsome face. "What do you think you're doing?" His voice was just as infected with anger and tension in every word he spoke. Finally, he took several threatening steps toward me as I said, "Making you angry, so you'll wake up."

"Oh," he said as my back hit his wooden wall. Damn it, of course. His arms locked me between him and the border with nowhere to go. He stared down at me as he said, "So you're trying to piss me off?" His lips were surprisingly close to mine. No guy would ever be this close to me in real life, let alone me being able to handle it; I would've been so timid and afraid. And if I did speak... I wouldn't be able to say a word.

"Yeah, I am," I said. "What are you gonna do about it?" He paused as his eyes looked between the two of mine. This could go one of two ways... violently or... a little more pleasantly. Either way, I was ready.

He smirked as he swiftly closed the gap between our lips.

Okay. So, maybe I wasn't ready. His body was as warm as his lips pressed against mine, gentler than expected. It was strange how real it felt and how easily I managed to kiss him back, even though I'd never kissed anyone before. His hands slid around my waist and pulled me closer to him. I could've easily stayed distracted as his lips began to move with mine—it felt... nice. But my mind dragged me back toward my goal: I had to get out of this dream.

"Wake up," I reluctantly mumbled. The guy pulled away, confusion written across his face.

"What are you saying?" he asked.

"You have to wake up," I said. He stared at me blankly before leaning in again. So he was a little harder to awake. No biggie.

I dreamt up a large hole behind him, one of my favorite alternate methods to waking people up when my words failed. I shoved him away from me as hard as possible, and he fell right in. The tug of wakefulness took hold of my mind as I snapped awake.

It took me a minute to realize I was staring up at the dark ceiling above me. I turned on my lamp and quickly scribbled down the dream and what he looked like. I re-read it several times just to feel the intensity of it again. Eventually, I shut off my lamp and rolled onto my stomach with a satisfied smile, drifting back to sleep.

***

It was 9:00 pm the next day, and I had just finished my schoolwork. I was absorbed in a good Wattpad read until a message pulled me back to reality: You coming to the party tonight?

Oh. Great.

Are we really welcome? I responded to Mindy. What if it's a trap or something?

I could sense her rolling her eyes. You're such a fool ofc we are! The most hated clique in the school is invited and bringing booze.

Of course, she'd be serious about going—why wouldn't she be serious?!

I tried in vain to escape. Ehh... not really my scene...

Come on!!! The whole gang is going. It'll be fun. I'll pick you up in 20! No discussion!

I sighed. Mindy was so stubborn, and I wasn't persistent enough. I climbed out of my cozy bed in search of my mom. Maybe she'd say no, and I could stay home.

The silence of our house was louder than ever, and I couldn't find her at all. Noticing that I was at a loss in the kitchen, Timmy walked up to me and licked my hand. I smiled and rubbed his head. My phone then buzzed.

Baby, I'm out on a date. There's food in the fridge. If you plan on going out tonight, let. Me. KNOW!!! Love you, my mom texted.

Darn her, I thought.

I then replied, Mindy is taking me out to a party. I went to get dressed, and Timmy followed with a toy in his mouth. I put on an oversized beige sweatshirt with black jeans and ankle boots in my attempt to dress fashionably. As I sat down on my bed to get more reading done, a car horn sounded outside, signaling it was time to go.

"You ready?!" Mindy asked as I hopped into the car. "'Cause I'm so excited!" She began blasting music as she drove down the road, singing along to the song. The night stretched out around us, and although I was being dragged off into the unknown of the party scene, I was glad to have Mindy there next to me.

Yikes. That statement was a little more dramatic than it needed to be, but... in my defense, I felt like my heart was gonna burst out of my chest! Her bubbling energy balanced out mine.

We knew we were at Grey's house when we saw it was surrounded by cars and could hear music blasting from inside. The place looked oddly familiar, but I couldn't put my finger on where I'd seen it before; I didn't venture into the ritzy part of our suburb, so it was beyond me.

"Whoa," Mindy said. "Look at that crowd." There was a sea of people inside the house, more than outside. I pulled out my phone and typed out, Is the rest of the squad here already?

"Well, yes and no," she said thoughtlessly, "Blake and Anastasia are out on a date..." Her eyes widened, realizing the oopsie she had made. Blake and Anastasia, who were my closest friends, didn't even tell me they liked each other, let alone that they were going on a date! I felt my heart start pounding in anger and pain. Why wouldn't they tell me?

I think she saw the hurt on my face because she said, "Who needs 'em? Let's have fun tonight! Everybody else is here."

My mood was soured, but I still nodded. Mindy started up the car again and wrapped around the street, parking a little further away from the house. We walked towards the mansion, past the rows of already drunk teens making out, laying on the grass, or both. I'm surprised there was no police activity by now. But, then again, the rich look after the rich, don't they?

The house was ginormous. A grand gate was left open, along with a cobblestone road that diverged into a circle around a mini gated garden. We walked up the hill and into the house, whose doors were wide open for the guests to enter and exit as they pleased.

I realized I slightly recognized the interior while walking in, which struck me uneasily. Why did I know this place when I had never met Grey in my entire life? As we continued making our way inside, music was pumped throughout the halls, leading us to the room that was the most active. Then we heard, "All right! Let's have some fun! You, you, and you get on stage and dance for us!" As we approached, Arnold bumped into us as he hurried out—he looked panicked.

"What's wrong?" Mindy asked.

"Grey's choosing random people to start dancing on stage," he said. "I wasn't gonna make a fool outta myself—no way José."

I stepped back with Arnold. Oh dear god, I'd rather die than set—

"Oh, come on, you're a bunch of babies," Mindy teased, pulling us both towards the room. "Just let me get a look at Grey, and then we can hang out in the halls or something like the losers we are."

We meshed into the crowd of drunk, rowdy teenagers to see a guy on the platform with his toned back facing us. He looked like he was speaking to the DJ on stage as several girls and guys danced on the platform rather lewdly. The crowd liked it, judging from the sound of their cheers.

"All right, who's next?" Grey said, microphone in hand, turning around. My eyes widened, and I immediately recognized him. It was the guy I kissed in the dream! I kissed Grey?! Well, technically, I didn't... but I kissed Grey?!

"You, in the green sweater... you with the Minecraft shirt and yo u..." His eyes scanned in our direction, and I immediately ducked. "You with the short straight hair." Mindy immediately whooped and pushed her way through the crowd and stairs. I sighed in relief, and Arnold stepped next to me, laughing.

"At least she's getting what she wants," he said. We watched as she danced, trying hard to make her moves enticing to Grey. Finally, he bought into it because he pulled her close to him and followed her movements, keeping pace with her hips and running his hands along her body. I cringed.

I felt a tap on my shoulder. "Hey, guys!" My eyes met Andy's smiling face. I waved at him, and Arnold cut to the chase. "Look at Mindy!" Andy's head snapped in her direction.

"Oh god!" he exclaimed with a mortified expression. "We can't take her anywhere."

After a few minutes of dancing, Grey pulled away from Mindy and went to the mic again. I pushed Andy in front of me to hide behind him.

"Who are my next victims?" he asked as Mindy and the rest of the people got down from the stage. She pushed her way towards us. I pulled Arnold next to Andy to make a double barrier.

"No, you should be hiding me!" Arnold said with a terrified look as he peered at me from over his shoulder.

"You, in the bright yellow crop top, you in the gray hoodie and..."

"Stop, let me go," Arnold whispered, trying to move away.

"Calm down and block her, dude," Andy hissed.

"You, in the black and you in the blue! Who's hiding behind you?"

# CHAPTER 3

Andy shrugged, and Arnold stared, paralyzed. "To whoever's hiding, it's a party. We're all just here to dance and have fun." Instant anxiety overtook my body, and I squeezed my eyes shut in fear. A clammy hand gripped mine and attempted to drag me from behind my impenetrable human wall. I yanked my arm back, and the guy rolled his eyes, pointing at me to confirm I was behind them.

Friggin jerk!

"Well, either they come out and come up, or one of you takes their place," Grey said.

"I gotcha, Hazel," Andy said nonchalantly. I moved behind Arnold, who slumped, relief overtaking his body.

I couldn't stay. Grey would probably want me to go up or have Arnold take my place. Arnold would die if that happened. So once

Grey's attention was fixed on everybody on stage, I yanked Arnold out of the room with me.

"Thank you," he said, relieved. "Sorry about that in there... I was really—I'm just sorry." I waved it away and patted his back.

We waited there together and talked, me using my phone to respond to what he said. "All right, everybody, get off my stage." A few minutes later, Mindy and Andy came strolling out of the room.

"You guys survived!" Arnold said. Andy pulled him into a headlock and gave him a noogie.

"You were so annoying," he said, laughing.

"Nu-uh!"

Mindy placed her hands on her hips and playfully glared at me. "And you! How dare you make Andy go up instead of dancing with that beautiful specimen?"

Gorgeous, actually.

"We're gonna have a quick break," Grey said over the microphone. "Get some drinks and stuff. That's what I'm gonna do." He sounded so casual. It was weird that he could be that way right after such a bad breakup.

As people began to file out, I grabbed Mindy by the arm and started to tug her down the hall.

"What?" she asked and laughed. I didn't want her to see my panic, but I was freaking out. Grey could not see me. If he recognized me, my secret could get out, and everybody would think I was insane—creepy even. It jumped the gun a little, but I never wanted to risk it.

"What's up, guys?" I heard Grey's voice shockingly close behind me. How the hell did he dip through the crowd so quickly?! I tensed up, my eyes meeting Mindy's. She gave me a confused smile.

"How's everybody enjoying the party?"

"It's really fun!" Mindy said.

"Yeah, it's cool," Arnold added.

"How about you?" he tapped my shoulder. I lightly flinched.

"She doesn't talk," Mindy said. "She signs, though."

"That's perfect; I love practicing my ASL," Grey said. Mindy smiled widely, then spun me around to face him.

His laid-back expression sharpened with recognition; my stomach dropped, dread pooling into it.

"Do I know you?" he signed. I shook my head a little too quickly. He narrowed his eyes as he seemingly tried to remember.

"We've never met before," I signed nervously.

I could tell he didn't believe me, but he didn't push it further. Instead, he shrugged and walked off.

"What was that?" Andy asked. I shook my head, unable to do anything else.

"Come on, let's go follow him," Mindy said, hitting me on the back. "This is the only night we'll really be intermingling with them. We gotta break down some walls and get on their radar!" I shook my head frantically.

"Yep!" she said. "And we're gonna take him up on his offer and get some drinks." She dragged me into the kitchen where Grey and some of his buddies were drinking beers.

Grey's eyes locked onto mine.

"Hi," Andy said. "Dancing on stage earlier was fun." He shifted uncomfortably as all eyes fell on him.

"I guess, " Grey answered, face leisurely turning towards Andy and breaking eye contact with me. "Since you liked it so much, why not be the first person on stage when I do it again?" Andy's fingers twitched.

"Nah, that's okay, " he said. "Thanks for the offer, though."

"I enjoyed it so much," Mindy butted in eagerly. "I'll do it again."

"You weren't half-bad up there," Grey commented. "I wouldn't mind." His attention returned to me. Shivers ran up my spine.

"Have a beer," he offered. I shook my head.

"Why not?"

"I don't want it," I signed.

"Do you have to be such a drag?" he signed back. My eyes widened in surprise.

"All because I don't want to drink?"

He smirked. "You came to the party, and you're not going to drink? That's pretty lame."

"You're lame for trying to pressure me into drinking." Then, some of his friends started snickering at our interaction—probably laughing at his and mine facial reactions; ASL is definitely an expressive language. The thought of them making fun of us—mostly me—made me angrier.

"No pressure. Merely pointing out the stupidity of coming to a party only to be a killjoy."

Annoyance bubbled within me. "Shut up."

"Why don't you make me?"

"No wonder your girlfriend cheated on—" I regretted it as soon as I began to sign it.

"Oh, yeah?" he asked. "You don't want to finish that sentence?" An awful, awkward silence overtook the room. He looked pissed but quickly changed his face into a calm gaze. My eyes latched onto the floor as he steadily approached me. I felt frozen—every muscle in my body shook with anticipation.

His breath tickled my ear as he whispered, "You're fortunate enough to be pretty, so I'll let that slide. But, if you ever mention my ex again, I'll make you regret it." He stalked out of the room, and his friends followed.

Mindy let out the breath she was holding in. "What did you say?!"

I shook my head, exhaling a breath I hadn't realized I was holding in.

"Time for round two!" I heard Grey's voice announce. "Who's ready?" I rolled my eyes. I looked at Mindy, expecting her to be jumping for joy and pulling us over to the room. But instead, she looked annoyed.

"You guys want to get out of here?" she asked. I nodded so vigorously that my head protested in pain.

"I refuse to stay here while he intimidates you," she said, and warmth filled my heart. We walked out of the loud, alcohol reeking house and onto the freshly cut grass outside. The scent was soothing to the nose.

"Wanna meet at DQ's and get something there?" Mindy asked Arnold and Andy.

"Hell, yes!" Arnold said. "The only good idea I've heard tonight."

***

"The way he moved with my body," Mindy said excitedly, "was so sexy. You could feel his muscles through his shirt!" After that, she hardly focused on the road as she drove me back home from DQ. We hung out there for a bit, eating ice cream before we parted ways, Mindy driving me home and couldn't stop talking about Grey. Fortunately, she was sober. I couldn't imagine how she'd be if she gushed over him while tipsy.

I got home relatively early. I didn't expect to, but we were all tired. We actually put in the effort at school, unlike half those people at that party who didn't give a damn, so we wanted to get sleep.

The car came to a halt in front of my house. The lights were on, so I assumed my mom was home, watching TV.

"I'm never gonna get over this night," Mindy said. "Thanks for coming with me." I nodded and smiled, giving her a hug. She hugged back tightly, squeezing the air out of me. "See you on Monday!" I got out of the car and slammed it shut, the breeze around me giving me goosebumps. I waved at her and then rushed into the house.

"Oh, baby, you're home!" my mom said, surprise in her voice. "I didn't expect you this early." A tall man held my mom in his arms. They stood in front of the TV, dancing to its music. There was wine on the coffee table, and two glasses had been poured.

He had lean muscles, a jawline sharp like a blade that made me think he'd be able to stab somebody with it if he so pleased to, bluish-gray eyes, and wavy brown hair; I wanted to punch him. My dad was 10 times more attractive than he was.

What the heck were they doing anyway? Dancing? The sight made me sore, but judging from the hopeful look in my mom's eyes, I knew she needed me to like him.

"That's the guy you met?" I signed, although I knew the answer. She nodded and swallowed. Approaching them, I offered my hand out to shake. His hand firmly clasped mine, and he dragged me into a hug. I internally screamed as I gently hugged him back. He let go with a big smile on his face. "It's nice to meet you, Hazel. I'm Jared." Cool, but I don't remember asking you for your name, heathen!

"Tell him I said 'nice to meet you too,'" I signed, trying not to look too hasty to get away from him.

"She said it's nice to meet you," my mom said, smiling. I could tell she was so excited that we finally met, but I needed to get away from them. I didn't want to be a party pooper, but this was the last thing I wanted to come home to. "I'm going to bed now. Goodnight."

"Goodnight, baby," my mom said.

"Nice meeting you again, Hazel."

I couldn't leave fast enough. Timmy chased after me, thinking that my run was a game. Well, it was far from it.

I collapsed onto my bed with a heavy sigh. Two things cluttered my mind: that thing with my mom and then Grey. First, I wished I had never gone to that party; now, I was on his radar, and second, my mom let this guy into the house in my anticipated absence.

My go-to person to vent was Anastasia. But, as I was about to hit send on a thoroughly fleshed-out message, the remembrance of her withholding information reared its ugly head. Frustration cluttered my chest, so I threw my phone across my bed, curling up into a ball. I didn't know how to feel.

It wasn't the end of the world—I at least knew that for sure. But with all my crazy hormones running around, it sure felt like it.

I got up and showered. When I came back out, I still heard the music playing downstairs. I didn't even want to picture what could be happening. I laid down on the bed, Timmy curled up next to me, and we slowly drifted off to sleep.

***

Discomfort filled my chest.

Please... don't let me be in Grey's dream.

I walked down a slightly narrow forest path. Trees encased me on all sides, the wind creating haunting whispers through the branches and leaves. Monstrous barks echoed in the distance, adding to the ominous atmosphere. Although the sound was terrifying, I knew not to be afraid. I was a lucid dreamer and in complete control.

Sometimes I had to remind myself that.

The barks got louder, along with someone's labored breathing. In the distance, a figure bounded towards me, its run sloppy and disheveled. It was like it was drunk. As it got closer, I could see it nice and clear.

Grey. Of course, it was him.

As he got even closer, I could see the fear in his eyes. I peered over his shoulder and saw that a large, black wolf was close on his tail.

Interesting.

"Run!" he shouted. I ignored him as he zipped past me.

I broadened my shoulders and legs as the wolf slid itself to a stop. A snarl came from deep in the back of its throat, white froth leaking out of its mouth.

I had to give Grey credit—he had one heck of an imagination.

"Sit!" I ordered. It complied immediately. Ah, I loved lucid dreaming so much. "Good boy!" I wrapped my arms around him as he started thumping his tail on the ground. "You're such a good boy." The wolf rolled itself on the ground, belly up.

"How'd you do that?" Grey asked, suddenly behind me.

"I just can," I said. I faced him. He was close, and his body heat emanated off of him.

"You're the pretty girl I saw at the party," he acknowledged. "I dreamt about you before." My lips sealed to keep in my fluttering heart.

"Why are you here?" His paranoia was readily detected in his voice, and I realized he was still in this nightmarish landscape. Every tree rustling still sounded like whispers, and they were all dark gray with black leaves. The sky held a menacing, stormy atmosphere. Even the wolf growled at him, though it stayed belly up on the floor.

My hand, palm up, reached for the sky. The sky cleared with a swiping motion, and a vibrant sun brightened up the atmosphere. Everything kissed by its rays was beautified; the trees bloomed with rich colored leaves, and their bark turned back to its natural auburn color.

Timmy's wet nose nudged my hand, replacing what was once the formidable wolf. I smiled at my work, satisfied.

"Hey," Grey said. "What are you doing here?" he asked again. His voice was friendlier now than I'd ever heard it before. My eyes met his.

"I come out here for walks with my dog." On cue, Timmy barked behind me.

"Can I join?" he asked.

"Sure." He fell in pace with me as random people having picnics phased between the trees. His brain was setting the scene for us.

Who was this guy?

"I'm afraid of dogs," he said suddenly. Timmy let out a friendly bark, walking next to him. That caused Grey to come reeling into me.

"Look," I said as I knelt down on the ground. Timmy licked my face. "If you're afraid, they sense it, and it throws them off. So the less fear you show, the better for both parties."

Grey's hesitated, but finally, he knelt down next to me. Then, warily, he held his hand out over Timmy.

"He won't bite." My fingers clasped around his palm, warm to the touch, and placed it on Timmy's head. His tail wagged, giving the most adorable puppy-dog eyes I've ever seen. A smile illuminated Grey's face, stunning me into a deep silence.

"You're a good boy," he said. He kept complimenting Timmy, making him fall onto his back, tummy up. "Dogs aren't too bad, I guess." I nodded even though he wasn't looking at me. He squatted down to rub Timmy's stomach. "Thank you for helping me." His eyes met mine as I knelt down beside him.

"No problem." We stared at each other for a moment, trying to think of the right words to say. I was definitely interested in getting to know more about him. I wonder if he felt the same way about me?

He opened his mouth, only to be interrupted.

"Grey," a recognizable female voice said from behind me. Startled, I spun around and immediately lost balance. Grey caught me as we ended and fell to the ground. My back was against his chest, and his legs were on either side of mine.

Sasha stood in front of us. Her hazel eyes were pooled with tears as her blond hair sat in a bun behind her head. "I'm so sorry. Please forgive me."

The sky around us turned a bloody red, and control of the scenery slipped through my fingers. Grey's grip on me tightened as he stared up at her.

"It's too late for apologies," he said darkly. "You hurt me, Sasha."

"Evan was a mistake," she said, sinking down to the ground. "Please, forgive me."

A black cloud descended from the sky, panic filling me at its sight. Screams erupted from all around us, too vivid not to make me afraid. I had to get the hell out of this nightmare.

"You're dreaming!" I shouted. "Wake up!"

Immediately the dream faded back, and relief washed over me. Remind me to never go into one of Grey's dreams again, I thought. Once the fear of the nightmare left my system, I drifted back to sleep.

# CHAPTER 4

On Monday, I stepped into Homeroom with my eyes fastened to my shoes. I sat down at my usual table with Blake and Ana already waiting.

"Hey!" Ana said cheerily. I nodded curtly. Immediately she sobered up and shot Blake a wary look. I fixated my attention on my phone before they could ask any questions. An ever-present tension danced around our table.

"Haze... something bothering you?" Blake asked. Yes, actually, but I can't tell you that because I wasn't supposed to know. But instead, I shook my head, desperately wanting to focus on the words from the story I was reading. Instead, my eyes skimmed the same sentence repeatedly as their eyes stayed on me.

"You're acting weird," Ana said pointedly. "Something's up. What is it?"

I wouldn't tell them; they didn't want to let me know they had feelings for each other. So I was going to be upset.

Our table sat in silence until the bell rang. I waved and then zipped out of the classroom. Eyes burned into the back of my head in the hall, causing me to spin around, expecting to see Ana catching up to me.

It wasn't her.

Grey strolled down the hall, this group of friends surrounding him. They were in conversations amongst themselves, but his eyes drilled holes into mine. Heat flooded into my cheeks and my eyes watered from the heat. I quickly returned my focus forward and navigated the sea of miserable high school students. Even in the crowd, I could still feel Grey behind me.

Andy stood in the doorway of math, and I was instantly relieved to see him. I threw myself into his arms; he clumsily caught me.

"You okay?" he asked, surprise laced into his voice. I nodded, too tense to do anything but that, and waited for Grey to pass by. He ignored us. "Was that jerk bothering you?"

I removed myself from him, shaking my head. My eyes chased Grey's form as it disappeared into his class down the hall. I reluctantly pulled out my phone and typed, I'm just a little nervous around him after the party. But, it was more than that. I was worried because he remembered that he had a dream about me. And then we were in another dream together. I was in dangerous new territory.

Andy searched my face. "What the heck did he say that it made you so nervous?"

"Ahem, enough with the lovey display; I want to start class," Mr. Johnson said. We both blushed with shame and hurried to our seats. I hope nobody heard our conversation. How long had the class died down?

I could hardly focus. Grey, Ana, and Blake robbed me of my focus. Class stretched out at an agonizingly slow pace but was still over too quickly. I slowly packed my things and headed toward the door. Andy caught up with me, escorting me in silence to my next class. I nodded appreciatively, which he returned with a smile before I entered English. Unfortunately, he had Spanish, so he disappeared down the hall with a mass of other students.

Mindy already sat in our corner, bag as a placeholder for me. She looked up from her phone as I got close with an expectant smile.

"Well, if it isn't my favorite person," she teased as she moved her bag. I rolled my eyes and slid into the seat next to her.

I searched through my bag for my Macbeth book when a familiar blond head bounded into the room. Her hazel eyes no longer had tears ready to overflow, and her pin-straight hair fell past her shoulders, bursting onto her backpack.

Sasha. She was gorgeous, but not because she was blond—that disclaimer needs to be put out there. I swear you could dye her hair green or shave her head for all I care, and she'd still be stunning.

I scoffed. Why wouldn't Grey have been dating somebody like her? She possessed the looks, and her figure wasn't horrendous either. It just proves how shallow he is. She probably didn't even have a fantastic personality.

Whoa there, that sounded mighty jealous. I shook the thought away. I didn't focus much on English as well. After class, I attempted to escape the room with Mindy. I wanted to get as far away from Sasha and out of the halls before I could potentially see Grey again. Unfortunately, my wish was short-lived as I bumped into him right outside the door.

"Well, what a pleasant surprise," he said, smugness and sarcasm hugging onto his words like their nonexistent lives depended on it.

"What do you want?" I signed angrily.

"Whoa, chill." He laughed, gently moving me aside. I turned and saw Sasha with an innocent and surprised look.

I wasn't surprised. It was that dream. When Grey saw her cry, he felt slightly moved even though he was angry.

"Are they still dating?" Mindy whispered as he made his way into the room. I shrugged, but I was sure we would find out soon enough.

***

"Hazel, are you mad at us?" Blake asked as he sat down beside me. My guilt prompted my response to his question.

"No, just girl things have me acting a bit crazy," I signed. Blake lit up at my response, which caused Ana to whisper, "PMS?"

I nodded.

"Why didn't you just say that?" she asked, engulfing me in a hug. "I thought we did something wrong." We sat at our usual table for lunch. Andy and Arnold still hadn't shown up, but none of us had the patience to wait for them as we tore into our food. They again asked me what was wrong. I knew I couldn't keep giving them the silent treatment.

I was angry, though. But it was primarily irrational anger. I'm sure there was a good reason why Blake and Ana wouldn't tell me about their feelings. I'd have to wait until they were comfortable enough to share. And even though I felt guilty lying to them, at least Mindy wouldn't be catching any heat from them for slipping up.

Sasha and Grey walked into the cafeteria together. His face was flush with anger, and Sasha looked smug, even though I could tell she was trying to show otherwise. The two of them side by side was enough to get the whole cafeteria in a trance as they approached their usual table. Grey slammed his hand down where his buddies were, Evan included. His eyes went wide as Grey harshly spewed out words that weren't loud enough to hear. Those who were close to them stared at Evan, horrified.

The chatter around us began to die down.

"That's such a lie!" Evan shouted and then pointed at Sasha with a shocked laugh. "You...! You lying whore!" Grey lunged for him. His fist clenched tightly to Evan's shirt.

"Watch your mouth," he said, his voice deeper than imaginable. Evan's eyes were wide as he raised his hands in defeat, and Grey harshly shoved him back into his seat. Together, they sat down at the table, Grey's arms wrapped around Sasha's slender shoulders. Their backs were to us, so we could see Evan's angry face as Grey ordered, "Get out of here." He stood, jerking his backpack over his back. He picked up his tray and lingered for a moment. We all went still and watched with bated breath.

In a flash, Evan threw all his remaining food onto Sasha. She snarled. Evan booked it out of there before Grey could even process what happened. Sniggers erupted around the room from various tables. Grey gave them one death stare, and that shut them up. My eyes were too attached to the scene to look away for my good. Grey's eyes met mine, challenging and hard. He was so unrecognizable from the Grey in his dream yesterday.

"I'm sorry that happened," I signed and then faced my friends again before I could see his reaction or response. We all looked down at our food as if it were the most exciting thing on the planet. Mindy's body brewed with unspoken energy, her leg jittering excitedly from underneath the table.

Slowly, whispers turned into full-on conversations, and the cafeteria came to life again.

"Holy crap!" Mindy exclaimed. "That was wild!"

"I wonder what that was all about?" Blake whispered, peeking over his shoulder.

"Whatever it was, it's best not to get involved," Ana said, steering his attention back to the table with her fingertips. It was evident that they liked each other, and I can't believe I had been so blind to it before.

I don't think either of them fully processed what she did—or maybe they didn't care—because they chomped down their food as if nothing happened.

Mindy glanced at me from across the table, and I just shrugged as if it didn't bother me. It didn't—not really. I'm happy for them, just not satisfied that they haven't told me.

Andy and Arnold walked in, grabbed some lunch, and then sat next to Mindy.

"What's up, guys?" Arnold asked.

"You missed everything," Mindy whispered excitedly. "I'll tell you once they pass." Andy and Arnold glanced behind us, causing me to do the same.

Sasha walked with her head high, even though her clothes were drenched in ketchup and grease. Grey walked beside her, carrying their backpacks on his shoulders. He paused right behind us, and I spun away from him nervously.

His breath tickled my ear a few seconds later as he whispered, "I don't need your pity, Mutey."

Something clicked deep inside me. Grey's presence disappeared, and for a moment, I sat there, seething in anger. Did he call me Mutey?

Mutey, out of all things! He insulted who I was because he thought I was pitying him?

Grey was covered in juice before I even processed what I was doing. It dripped down their bags and his back. He had walked through the double doors of the cafeteria right as the cup made contact. Considering I hadn't planned a direct hit, it was satisfying how great my aim was.

I spun back around and faced my table, all of their eyes wide as saucers. They looked at me as if I was a stranger on the street, acting crazy. Pissed off by their reactions, I pulled out my phone and wrote, HE CALLED ME MUTEY!!!

Andy and Blake were behind me with lightning speed. I turned and saw them blocking the path of Grey, his eyes surprisingly calm. I don't know how he did it.

"Don't make more of a scene," Grey said. "I'm going to sign to her."

Andy took half a step to the side, still partially blocking me. "Then do it."

Grey looked me square in the eyes and signed, "Good aim. You'll regret it."

I smiled cheekily and responded, "There's no way I'll regret something you deserved." Slight confusion and annoyance flashed across his face, and he walked off back to Sasha, who was waiting outside the doors. She gave me a dirty look before they disappeared down the hallway.

***

"So, that was a little crazy," Ana said, patting my back. We all sat in my room, trying to get homework done after school that day. We all lost that focus rather quickly. I shrugged.

"What did he say to you?" Andy asked, worried. I shook my head.

"Come on; we're all concerned here," Andy pleaded.

"Do you want one of us to tell them?" Blake asked. Both he and Ana saw what he said.

I sighed and pulled out my phone, typing, He said I would regret throwing the juice at him.

"That's not good," Mindy said. "These popular kids... if you get on their bad side, they go for your blood." Pft, okay.

"Yeah, no one will hear you scream," Ana teased, shoving me lightly. I rolled my eyes, knowing that wasn't true. I'm still capable of making noises, just incapable of speech.

I returned my attention to the precalculus homework I needed to get done. My friends returned their attention to their work, earning the room a blissful silence.

After at least half an hour with mini conversations here and there, Timmy started barking from downstairs. My mom had to be coming home.

"You think you should tell your mom?" Ana asked. "She can make teachers be on the lookout?" I shook my head. It was honestly fine.

Grey and Sasha were all talk, no bite.

# CHAPTER 5

That night, Grey dragged me into another one of his dreams. Palm trees surrounded me, along with discomfort, as I sat down on the ground beside him. The sky was dazzling, but debris lay all around us. It was as if a tsunami had torn by here.

"Why are we here?" I asked.

"This is where she drowned," Grey said, leaning his back into a tree. My throat tightened as I glanced around me. "She drowned in the tsunami."

"Who?" I whispered. His eyes watered, but not a single tear touched his face. Instead, they suddenly gazed into my eyes. "Sasha."

I expected her to be behind me, but she wasn't. So I returned my focus to Grey, and his soft lips suddenly encased mine.

Surprised, I shoved him away. "What... what are you doing?"

"What do you mean, Sasha?" he asked. His dream was a bit erratic that night. One moment he's talking about someone who drowned; the next minute, he's thinking about Sasha.

"I'm not Sasha," I said slowly. "I'm Hazel." His eyes narrowed, and then he leaned forward. "You threw juice at me." There we go. Grey was back.

"You called me Mutey."

"So?" he asked. "Me giving you that nickname justifies it?"

"Yes."

"So you think you're allowed to humiliate me, and I should just take it?"

"You think you're allowed to insult me, and I should take it?"

"Insult you?" He looked at me angrily, even though he was the one who caused bad blood between us. "I haven't done anything to you, Hazel. You've started with me." I rolled my eyes and then imagined a hole where the tree once was, not even bothering to tell him to wake up.

"Sure I did," I said with a humorless laugh. "In you go."

***

Surprisingly, the week passed without incident and without me getting dragged into any more dreams with Grey. He was my least favorite person at the moment and the last person I wanted to share a dream with anyway.

"You look tired," Ana said as I forked through my lunch that Friday. I looked up and shook my head, offering her a smile. "Seriously, what's wrong?"

"Nothing," I signed and meant it. I didn't know where Ana's comments were coming from, and I was slightly frustrated, signing, "I'm fine." I could tell she didn't believe me. Everybody at our table looked at me like a puzzle with a missing piece. "Honestly!"

Grey said, "Everybody, shut up!" We all looked and saw him standing at the table. I immediately cringed in horror, knowing someone would eat on it. That would be the first thing you thought of, you fool. I laughed at myself internally. "I'm having another party at my house tomorrow. Juniors and seniors are invited to join us." I rolled my eyes and faced away from him.

"Get off the table," one of the teachers shouted from across the room. Grey hopped down nonchalantly.

"We have to go," Mindy said as everybody resumed talking. "We'd be lame if we didn't."

"I'm not going," I signed. "I don't want to be around Grey."

"She said she's not going," Blake translated.

"You have to go," Mindy whined. "I need my party buddy!"

"Your party buddy will be home, getting work done," I signed as Blake translated.

"I'll change your mind, my sweet, I promise."

"She doesn't want to go," Andy insisted. "Don't pressure her." She winked at me in response. She and I both knew that I was going to end up going.

***

That night, I had the nightmare.

We stood in the Mobile gas station, right outside the store. My dad went stumbling away from me as I was locked in the confines of an unknown man. I nearly wet myself as something pressed against my temple; I knew it was a gun.

"Give me the money right now," the masked man said, gun pointed at my head, "or I'll kill her." I looked up at my father and his helpless eyes as he quickly dug into his wallet.

"Faster!" the man ordered.

"Daddy," I said, tears escaping my eyes.

"It's okay, baby, I'm going to give the nice man some money, and we're going."

My father kept digging in his wallet without producing any money, which caused the robber cock his gun. "You have no money on you?" he asked.

"No, I'm so sorry; if we could just get to a bank or something, I'll get you the money," my dad said quickly. "I just have to get my card."

"Not necessary." The gun went off before I could process it, and I screamed and covered my ears. My father crumpled down before me. Droplets of warm liquid soaked my face, and I shut my eyes.

I couldn't speak anymore. The man fled without another word, leaving me in the cold night air with no voice and a dead father.

I woke up, sweat drenching my body and tears escaping my eyes. I bit my lip and closed them, willing the harsh memory away. I didn't want to think about it anymore—I had gotten so good at forgetting it, but it seems right when I think I've made some progress, that dream returns.

My dog stirred in bed next to me, his tail wagging as I sat up. I gave him a reassuring pat before walking off into the bathroom.

Water-soaked my skin as I splashed it onto my face. Its coolness refreshed me, the last thing I wanted at the moment, being that I was so tired but was needed all the same. The sting of the dream and memory was already fading away due to my waking up.

Gunfire rang out again as my dad crumpled to the floor. I shook away the image, not even having enough energy to cry. I had cried so many times over it, and it never made a difference; the robber took my father's life, and he'd never come back.

The cops tried hard to communicate with me at five years old, asking me what he looked like if I could remember the color of his eyes, hair, and skin. But I couldn't answer. The more they pressed, the more distressed I got. My mom finally gave up for the sake of my mental health. Once they finally got the CCTV footage, they realized it would've been impossible for me to see him. He wore a mask and sunglasses.

The case was hopeless.

I went back into my room, tears threatening to spill as my chest ached with pain, but I sucked them back in. Tears wouldn't bring him back.

Nothing would.

He's dead.

I finally wanted Grey to desperately drag me into one of his dreams because I couldn't think about my dad anymore. The thought of him burned me. It's almost like I was less capable of handling it at 17 than five.

Finally, after laying back in bed, tossing and turning, I shut my eyes and got some peace.

# CHAPTER 6

As much as I didn't want to be at Grey's house, Mindy somehow managed to rope me back in.

"It's going to be fun," she said as she parked a few feet away. "Trust me. And, everybody is going to be here this time!"

Whoopie... I rolled my eyes.

We walked into the house, immediately filled with the stench of liquor and sweat. I don't know why I allowed her to bring me here again. Coming here wasn't a good idea; I could feel it.

There were never mobs of kids around the front entrance. The party always seemed to occur in that room with the makeshift stage. It was huge as well, so it could fit a sizable amount of the crowd.

"Andy said they're in the big party room. Let's go!" I followed Mindy, and we entered the loud, hectic space filled with drunk teenagers. The group was already pressed against a wall, yelling over the music.

"Hey, guys!" Mindy shouted, pulling me over to them. "Did we miss anything?"

"Besides people dancing and drinking, no, not much," Andy joked.

"Facts," Blake said. "But we have some things to be nervous about, according to Arnold."

"I'm serious! Last time he chose random people in the crowd to dance! I was terrified!" Arnold's wide eyes and horrified expression made me chuckle. He was the dorkiest out of all of us.

"Well, it seems like there's no dancing going on on stage tonight," Ana concluded. "Why don't we enjoy this party?"

"I'm not putting my guard down," Arnold said. "I won't be caught lackin'."

I was with Arnold on this. Grey hosting another party for every clique seemed a little sketchy. Especially after what I did to him and what he promised me in his dream. I was sure he was up to no good. And I stupidly walked into the lion's den.

"'You enjoying the party so far?" Andy asked me. I shrugged, not feeling any particular way about it yet.

"She would if she'd drink," Mindy said. "But she never wants to."

"And you're not going to pressure her to do so," Andy stated firmly.

"Of course not!" Mindy snapped. I could tell there was some underlying tension between them, so I pulled Mindy into the crowd of dancing people. We began dancing with each other, and I enjoyed myself, even with the mobs of drunk kids around. It made me feel less self-conscious about how I looked if I was being honest.

The song ended, and the music stopped playing. Everybody looked around, confused.

"Hello, everybody!" Sasha said, standing up on the stage. "Thank you all for coming. We have quite a show to put on tonight." There was even more confusion within the crowd. A show?

"As you may know, we have a mute girl in our school," she said, "and she decided to show her true colors and be super rude. She's written awful stuff about Grey and me on her social media pages, and honestly, enough's enough."

Mute girl? I was the only girl who didn't speak in our school. And I had written nothing about them on social media pages. I hardly even used social media! And the fact that she even said Mute Girl... hurt and anger shot through my body.

"I'll be posting the receipts on Instagram later tonight," Sasha said. "But for now, how about we read an excerpt from a poem of hers on Wattpad?"

My stomach sank in dread and blistering anger. I pushed through the crowd towards the door, only to be stopped by Grey. His face was impassive as he spun me around to face the stage. He had a vice-like grip on my shoulders, making it impossible to get away from him no matter how hard I struggled.

"Let go of her," Andy said as Sasha began reading my poem. It was the one about my father. I closed my eyes, my body starting to shake.

"Stand back before I beat the hell out of you," Grey warned. I opened my eyes and shook my head. Andy backed off.

"Daddy, I miss your voice," Sasha said dramatically. "You gave me life/ brought me up from my worst moments/ and now you're gone, and we're alone/ it's no longer the same without you at home." Then, she began laughing, stabbing me right in my chest. "Somebody's really kinky, isn't she?"

Anger flashed through me, and I tried to dash towards her right then and break her stupid straight little teeth. But, Grey's grip was relentless and unmovable.

"Daddy, I'm all grown up now./ You missed me blossom into a young woman/ My mind-expanding alongside my body/ writing, like you/ which is now my favorite hobby." Laughter erupted around the room. Tears of shame washed over my face and down my cheeks. I wiped them away quickly, even though they kept flowing out. "Was this poem written by a five-year-old? It's honestly trash, guys. Guess she needs a better hobby—"

"You think this is okay?!" Blake shouted, causing everybody to turn around and look at him. "Her father was murdered right in front of her. So you think it's 'cool' to make fun of her poem coping with her grief?!"

The room went quiet until Sasha said, "Hah! Well, maybe next time she shouldn't mess with us. She deserves it." She shrugged with an ugly smile on her face. "But since people are so sensitive, I think I proved my point. The party can resume. Remember to check my Instagram page later today, loves!" Then, she turned the music back on, and everybody slowly started dancing again.

Grey sighed as he let me go, and I shoved past him, running through his house until I got lost in one of his various rooms.

I collapsed onto the floor, my sobs shaking me and my chest aching. I was completely and utterly humiliated.

My phone began blowing up in my back pocket, but I ignored everything sent to me. These emotions cluttered my mind and body, and I just wanted to cry it all out. Then, theoretically, I'd be able to act rationally. Right? I didn't know. I was so fed up.

I missed my dad more than ever. I felt like he'd have something encouraging to say, more than the bland things I offered myself in my mind. But he was dead, and nearly a thousand students from two grades combined knew my poem and believed that I was bullying Sasha and Grey.

A knock on the door startled me, and I silenced my sobs, listening. Nobody on the other side of the door made a sound. Then, when I thought the coast was clear, I silently started crying again, burying my face in my arms.

The door twisted open slightly until Grey's face peeked in from behind it. My eyes hardened as I looked up at him, hatred—an unnatural emotion for me to harbor—building up in my chest. I stood to leave, but he blocked the doorway and closed it behind him.

"I'm sorry," he muttered. I narrowed my eyes at him. The audacity! He was sorry? After this, that's all he could say? Sorry?!

I stood there, unmoving. Grey looked like he expected me to say something, but I refused to move my hands from their fists at my sides. More tears dripped down my face, only making me angrier. I didn't want to cry in front of him.

He continued, "I... I didn't know. I didn't know she'd read that. But you messed with us. I warned you you'd regret it." His eyes held guilt, but that wasn't enough to cool me off. The hate I had built up for him boiled into a rage.

"Hazel, I understand you're upset, but you are who threw juice—"

"Because you insulted me!" I took an angry step toward him. "All I said was, 'I'm sorry!' What Evan did was crappy! But what do you do? Take what I said as me pitying you. And then you had to call me such a nasty name!"

"I didn't see it as an insult!" he shouted, causing me to take a step back. "I was being playful!"

"You insensitive bastard! Do you think I like being unable to talk? You think I want anyone to call me 'mute,' let alone—" There was no word for "Mutey" in sign language, which further frustrated me. I angrily clenched my fist together and quaked.

He ran his hands through his hair frustratedly. "I thought you were like... mute and proud or something."

"Proud?!" My eyes widened, and I snapped—an angry scream escaped my chest, making his eyes widen, and his mouth open slightly.

My hand met his face with a satisfying slap, the sound filling the room. He grabbed my wrist before I could do it again, pulling me against his chest. A flood of shivers swarmed my body as anger tore through my bones.

"Take a deep breath," he begged. "Please."

Tears escaped my eyes, soaking his shirt. Momentarily, I found comfort in his arms being around me. I wanted to lean in, press my face against his chest and let the emotions wash out of me.

But he was the one causing me pain. And I was stupid for letting him even try to make me feel better. I pushed away from him. His eyes left my face, unable to face the tears.

"I watched my dad get shot before me," I signed slowly. "His blood covered my face. His killer took away him and my voice with one tiny bullet. I couldn't say anything and still can't. And you made a joke of it in front of everybody."

He said nothing and swallowed hard.

"I hate you," I signed. "Let me out." He hesitated, his eyes lost. "Now!"

I managed to find my way back to the central part of the house. My friends sat outside on the lawn, waiting for me. All their phones were in hand, and my phone continuously vibrated in my back pocket.

I tapped Mindy, and she spun around, relief in her eyes. "I'm so sorry." She hugged me tightly as the rest of them got up and did the same. I graciously accepted their love and cried some more as they held me tightly.

"Go away," Andy said, pulling away from us. Ana and Mindy held onto me as all three of the guys stood before us. Grey looked at me through them and then back at them.

He looked like he wanted to apologize some more, but his eyes hardened. "You need to leave my party. All of you. You're not welcome anymore."

"As if we ever were," Andy snapped.

"Shut up, and get your girlfriend out of here." Grey's eyes were dangerously dark. Andy's fist clenched, but he knew it was a fight he couldn't win.

"Let's just go," Blake said. He pulled Andy back. Arnold was already behind us and pulling my shirt to lead us away.

"At least I'd have a girlfriend who wasn't a garbage human!" Andy snapped.

"What did you say?" Grey advanced toward Andy, who had a death wish. "Say something about her again; I dare you."

"About who? I didn't name anybody. It seems like you already know the trash you're with."

Grey's fist collided with Andy's jaw, and he stumbled backward. Mindy's surprised yelp and Grey's enraged body language caused people to come swarming around us.

"Fight, fight, fight!" a guy in the crowd shouted. Then, more taunting cheers started cheering for Grey, and Sasha appeared behind him wearing a smirk.

Andy knew he was in over his head. However, he still went for it, managing to punch Grey in his nose. That was all he could get in, though, before Grey kicked him hard in his stomach. He doubled over, groaning in pain, and fell in front of us. I pulled out of their grip and acted as a barrier between him and Grey.

"Oohhh, player three has entered the ring," Sasha commented in a sassy manner. "'You going to beat him up with your words?"

"He's done," I signed, staring at Grey. "He defended me and got his butt kicked. You've taught him a lesson. We're leaving."

"What did she say?" Sasha demanded.

"She said they're leaving," Grey said. "As they should."

"I think you should teach her a lesson for messing with us," she said. "Bruise that pretty face."

Anger flared in my stomach, and I hoped she'd take up that offer and come into the circle herself.

"We've done enough," Grey snapped, pushing through the crowd away from her. She raised her hand in mock surrender while letting out a piercing, sarcastic laugh.

People around us mumbled uncomfortably or booed and began to disperse. Some laughed and drunkenly swayed away. Sasha stood rooted in place as the crowd dissipated until it was just my friend group and herself. Everybody in our group bent down to help Andy as I kept my eyes on her.

"This isn't a game you want to play, Hazel," she said as she got closer to me. "I will make you regret ever coming to this school." She took the final step into my space, and I tried not to flinch.

She spat on my face.

# Chapter 7

Although my mind told me to relax, my body instinctively lunged toward her, knocking her to the ground. She screamed, surprised, as she landed.

Her eyes were wide as she looked up at me. I wanted to be tough as hell at that moment, but with spit dribbling down my face and my brain torn in two directions, all I could do was stand over her, my eyes mirroring hers. I blinked and tried my hardest to regain composure.

"Come on," Blake said, quickly pulling me away. His voice turned into a whisper as he said, "We don't want this to get bigger than it already is going to be."

"You're dead, Hazel!" Sasha shouted. "As dead as your father." Anger tore through my body, filling every vein with its venom. I kept my mouth sealed shut and acted like it didn't bother me. She wanted it to; I knew it. It took everything in my body to keep moving and not give her the satisfaction of my fist busting her hideously gorgeous face.

I got into the car with Mindy, DQ bound, wiping spit off my face with the napkin she handed me. Anger still embedded itself into my core; my stomach clenched as shivers wracked my body, even though it was warm in the car.

The drive was utterly silent. I didn't know if this made me angrier or if I appreciated its presence. Either way, the emotions I was feeling were rooted deep. It was going to take a lot of crying to get it out.

When we finally pulled into the parking lot, I slammed the door shut as we walked into Dairy Queen. My body still shook as I stared up at the menu. My friends were too quiet.

In a flash, my anger dissipated. I couldn't handle it anymore. I ran out of DQ, my friends following. I sank to the ground as a bucket load of tears flew out of my eyes. Ana and Blake sat on either side of me and wrapped their arms around me tightly.

I couldn't say anything, but I wanted to say everything. Like, what the hell could we do now? What were they going to do to our friend group? Why was I suddenly getting picked on after three years of high school? Why did this hurt so bad?

Why did Grey get back with that witch?! And why did he show me pity?

***

I sniffled as the last of my tears washed out of me, and a needed calm replaced everything.

"You okay?" Blake asked. I nodded. A few long minutes later, we finally got some ice cream. After some ice cream and lots of talking, I finally started to feel better.

"We've got your back," Mindy said. "Don't worry." We sat at a table in the DQ, warmth filling the rooms from the heat and her words. I couldn't appreciate them enough.

"Sasha's jealous of you," Andy stated. "That's why she's mistreating you." I looked at him, confused, and he smiled at my naivety. "You're prettier than her. That's just a fact because your personality seals the deal. And Grey has taken an interest in you; it's so obvious. The way he looked at you when you guys first met was so weird. Anyway, I don't think he views you like someone to hook up with, but you have a tune in his heartstrings if that makes sense."

The dreams? Could it be that? Did he still remember things that we said and did in those dreams? Were they vivid? Or were they a snapshot, which was still enough for him to feel something?

You know what? Who cared? Because of him, I was humiliated and ashamed of myself. Screw him and any goodness he possessed.

Mindy looked down at her phone and closed her eyes with an annoyed sigh. Then, she slowly put her phone back under the table and looked up at the rest of us, who were all looking back at her.

"What is it?" Blake asked in dread.

"Check Sasha's Instagram..." she whispered.

My heart slammed against my rib cage, immediately pulled out my phone to check her Instagram page.

They were screenshots of supposed messages between her and me.

Her: Hey, I wanted to know if you could stop talking crap about Grey and me. It's really childish and unacceptable, especially bc we did nothing to you.

"Me": No.

Sasha: I don't understand what your problem is??? We've done absolutely nothing to you.

"Me": well, ur existence bothers me.

Sasha: wow. childish much? there's literally nothing I can do about that. Are you suggesting I kill myself or something?

"Me": If the boot fits

She was right. She did make me regret going to this school.

"This is so easily fixable," Mindy said, seeing the panicked look on my face. "Those messages don't exist; neither does she have your number. So don't freak out."

That didn't make me feel better. So many things started running through my mind.

Is she going to tell the principal, even though he couldn't care less? Is she going to try to get me expelled? What the hell do I do? I closed my eyes and slowed my breathing before my emotions spun out of control.

"Don't panic... we can resolve this." I glared at Blake.

"You think so?" I signed angrily. "Everybody in school has made fun of me! There's no resolving it. I'm done. Let them think what they want."

***

The following Monday, I stood outside my school with a sickening knot in my stomach. Although the people passing me rarely spared me a second glance, I still felt super nervous. I stayed away from social media the whole weekend and kept my nose buried in electronic and physical stories to keep my mind off of everything. I also had an exciting dream with Grey, one where I got to wake him up in the best way possible: making it into an intense nightmare.

The dream started with Grey running through the woods again. I heard the wolf howling as it chased after him. I looked around on the ground until I found what I wanted—a large black cloak that looked like something Death would wear. I eagerly put it on and waited.

It took a long time for Grey to appear. He was running like his legs were made of lead, which we've all experienced when running in nightmares. I stood in his way, and he stopped quickly. I pulled out a scythe from behind my back, and to make it creepier, I leaned my head to the side in an exaggerated position and began to limp toward him. I made sure he couldn't see my face. He said nothing as he turned around and tried to run back to where he came from. But the wolf stood there and bared his teeth at him.

To intensify the nightmare, I dreamt up two humanoid creatures so that all of us surrounded him. They were gray, bald, and crawled on all fours. He spun around in a circle and began to scream. Again, I felt my brain being dragged awake. Then, with a satisfied smile, I fell back to sleep.

But this was the real world. And I was nowhere near as intimidating as I was in the dream—and even then, I'm sure Grey woke up and realized how ridiculous it was to be scared by something so stupid. So regardless, what could I genuinely do against him?

Summoning what little courage I had, I walked to the entrance of my school. Still, hardly a person noticed me nor cared who I was. They all ignored me as they walked by. They listened to whatever was playing on their earphones or continued to enjoy the conversations they were having amongst themselves. I tried to be lowkey and slipped through the crowd. I grabbed what I needed out of my locker for the day and inconspicuously rushed to homeroom.

I sank into my chair with a mini sigh of relief. Ana and Blake gave me pity-filled smiles.

"No, this day is going well," I signed happily. "It's as if I didn't exist, like normal; I love it."

"Well... just expect someone to say something stupid," Blake warned. "You know the kids in our school love drama."

"Yeah, but that's mainly the ninth-graders," Ana said. "And Grey didn't invite them so..." Ana pursed her lips and glanced between us playfully. I smiled and signed, "She's right."

"I mean, yeah..." He looked around the room for a moment. "Just don't let your guard down." I knew he was right.

The bell rang too soon, and I headed off to English class. I tried not to think that Sasha would be there as I slowly walked over, but my anxiety wasn't even allowed to build up—Sasha stood in the doorway of class with her arms crossed.

"Hope you've come here to apologize," she said, her brows furrowed. I tried to push past her, but her other friends stood in the way behind her.

"Apologize," she demanded. "Say it." Say it? My teeth clenched. She smiled at my angry reaction.

"Unless I hear you say it, you're not getting into class," she said with a smug shrug.

"Come on, 'mute girl,'" one of her friends said. Sarah, I think her name was. "What's wrong? You look upset." She smiled deviously.

"Girls, what's going on here?" Ms. Williams asked as she stood behind me. I almost jumped at her sudden presence.

"Not much," Sasha said casually. "Just talking about Macbeth."

"Excellent. I can't wait to hear what you thought about the reading! Inside, everybody."

They parted the Red Sea for me to walk past. I hesitated before walking through them. Sasha stopped to put her sickeningly soft hand around my upper arm. She yanked me close to her as she whispered, "We'll talk later, Mutey."

# CHAPTER 8

The rest of the day, I tried my best to stay out of dodge, even going as far as eating lunch in my car. I was afraid Sasha would have something else in store for me if I stayed in the cafeteria. Ana and Mindy sat in the car with me and talked trash about Sasha and Grey as we ate together. They made me feel better about everything and reminded me that it wasn't the world's end.

I felt like I was finally getting a grip until Grey was leaning against my car, arms folded, after school. We stared at each other for a long moment, and the silence stretched between us. Then, finally, I took a deep breath in and, in false confidence, strode up to him. "Can I help you?" I signed. His very presence intimidated me, especially after his girlfriend's promise, but I tried not to let that show. Did my fingers shake when I signed to him? Oh, dear God.

"I've come to warn you, Hazel," he said, and his eyes narrowed. "Sasha gets what she wants. And she wants to ruin your life. Apologize, and you won't have to deal with her anymore."

Blood rushed into my ears. "For what?!"

"For pushing her!"

"Do you recall all she did to me up until I pushed her? And to top it off, she spat on me!"

Grey let out a heavy sigh and mumbled something under his breath before finally saying, "She spit on you?"

"Yes, so I shoved her. Now, move! I need to get home."

He hesitantly got off the car and stepped out of my way. I opened my car door and sat down. Before I could shut the door, he held it open and bent down to meet me eye to eye. "...get home safe, I guess." How dare he act like he even slightly cared about me? If he did, he wouldn't be helping her make my life miserable. The audacity of this man!

"Stay away from me," I signed and then slammed the door shut.

***

When I got home, I tried not to seem upset around Mom. I didn't want her to worry about me, especially when she appeared to be on cloud nine. I haven't seen her this happy in a while. When I got home, she was dancing by herself as she mopped the floors. She was ecstatic. Apart from being hurt to see her that way and knowing it was because of Jared, another part of me wanted her to be happy, no matter the reason.

I wrapped my arms around her torso from behind, and she let out a surprised scream. I laughed as she spun around. "Baby, you scared me!" She kissed my cheek. "How was school today?"

"It was all right," I signed. "Boring, as usual." Classes were, at least. "How was your day?"

"I took the day off from work to clean this house up a little bit," she said and let out a satisfied sigh. "I made some pretty good progress. I'm going to finish the rest this weekend, though. Want some Chinese food for dinner? I'm too pooped to cook."

"I'm always down for Chinese food."

My mom and I watched a movie together that night since I'd gotten a decent chunk of my homework done during study hall. I spent the rest of the night on assignments and then had a slight bit of downtime (I would've had more if I stopped getting distracted by Instagram) before falling asleep. Grey didn't take long to join me in the dream world.

"You make things so difficult!" he snapped. I hazily phased into his dream that night, chains wrapped around my wrist. I was in a dark dungeon room, Grey wearing armor, as if he was in medieval times, ready for battle. I looked down at myself and saw I had on a raggedy-looking dress with dull gray and white colors.

"Are you listening?" I looked up at his frustrated face.

"No," I said.

"Damn it, Hazel." He ran his hand through his hair, clearly upset. "She wants to kill you, and you still don't even care?!"

I casually started to release myself, but his following words caused the chains to tighten. "I don't know why I care so much about you." I looked up at him, startled. "There's no reason to. I know nothing about you, yet I can't help myself."

"Seems like a personal problem," I retorted. "For someone who cares so much about me, you sure don't act like it."

"I know!" He kneeled in front of me and grabbed my face in his hands. A mixture of repulsion and warmth flooded into my chest. "I don't want to care about you! Every time she does something to you, I feel guilty. But why should I?" We stared at each other before a shrill, poisonous voice filled the dungeon chamber.

"Grey!"

He spun around, and we saw her there, standing by the entrance to my cell. She didn't look at him in betrayal like I would've expected. Instead, she looked at him expectantly. "You said you were here to interrogate her."

"I was, honey," he said as his outfit changed from a knight's to a King's. (You have to love how quickly the brain changes things in dreams.)

"You call holding her face an interrogation?!" she snapped. "We did things your way, but you betrayed my trust; I'm ending this." She pulled out a dagger from her boot and slowly advanced toward us.

"No!" Grey shouted, his body acting as a barrier between her and me as he pulled out his dagger. "Don't touch her. She doesn't deserve this."

"She has to die! She's taken you from me!"

"She's done nothing!"

Evan suddenly appeared beside me, yanking me up in his arms and placing a dagger, cool in touch, against my neck. "Drop the knife, Grey, or she dies." His voice was sinister.

Grey slowly turned towards us, shock and rage in his eyes. His dagger clattered on the ground, echoing throughout the dungeon chamber.

"That's a good boy," Evan said tauntingly. "Here you go, babe." He shoved me towards Sasha. She leisurely approached me with a slight smile on her face. Grey couldn't meet my eyes as she ensured I could see every glossy white straight tooth. "This is for getting in the way." I didn't feel the knife, but I did fall to the ground, Grey's brain dramatically killing me off. "Say your farewells to her and your throne."

They walked out, leaving Grey and me alone.

"I'm so sorry," he said, kneeling next to me. "I let Sasha hurt you again."

"That's all you seem to be able to do," I said pointedly. Grey flinched.

"I don't know what you want me to do," he said, "I am her boyfriend. I have to have her back."

"Doesn't seem like she ever has yours," I said. My presence in his dream was slowly fading.

"No... she doesn't. She always wanted both me and Evan."

"Yeah... played like her favorite piece on the chessboard."

Everything faded to black.

***

"Don't let me fall!" Mindy exclaimed. I had to keep myself from laughing too hard as she hit the ground and bounced. The grade took a trip ice skating on Tuesday, renting out a whole rink for ourselves. Mindy, who's super uncoordinated, managed to fall five times in a minute alone.

"You're making us all look bad," Ana teased, helping her up as I gripped the ledge. My laughter was uncontrollable.

"You're evil, Hazel," Mindy said and laughed along with me.

I looked around, my eyes landing on Sasha and Grey. They leaned against the ledge on the other side of the rink, lost in whatever conversation they were having. They honestly looked happy. Her smile was vibrant and genuine, and his smirk made him irresistibly attractive.

I returned my focus to Ana and Mindy, Ana trying hard to stabilize Mindy.

"You guys ready to go?" Ana asked. Her eyes followed Blake as he whizzed past us with Andy and Arnold.

"Yeah, yeah, we're comin'," Mindy said. She began skating hesitantly, her arms out, her legs way too straight, and she was already wobbling again.

"Go," I signed to Ana. "I'll hold her."

"You sure?" she asked. I nodded and shooed her away. She smiled and then skated off.

I held my arm out for Mindy, and she graciously accepted it. "Thanks, you're a real doll." I rolled my eyes at her joke, and we finally started to make some progress. After a few minutes of struggling to keep Mindy up and her finally managing to get into the flow of things, Evan slid next to us.

"Hey, guys," he said. "Need help?" Several more guys from his popular friend group surrounded us. My gut started to twist uncomfortably.

"That'd be great," Mindy said. "I'm wobbling all over the place."

"I'll take her off your hands," Brad, one of the guys, said and linked arms with her. They pulled off, leaving me to skate side by side with Evan as his other friends dispersed.

"How fast can you go?" he asked. When I didn't answer, his eyes widened in realization. "Sorry... I keep forgetting you're mute. My bad." I shrugged it off. "You know, Sasha wouldn't hate you so much if you just apologized to her. Your texts were rude."

I quickly detoured off the rink as he called after me. People whizzed past me, and I ensured he'd have a difficult time if he even tried to follow.

I was so angry that I couldn't contain my wobbles as I made a beeline to the benches housed outside the rink. I can't believe how many people so stupidly believe her lies!

After a few minutes of throbbing feet and anger, Evan decided to join me. He sat down on the bench next to me with a heavy sigh. I tensed. "I want your help. Grey has something going on for you. And I have something going on for Sasha. She can't have both of us, and she stupidly picked the wrong guy. From the way you were looking at him a few minutes ago, I saw that you have something of your own for him. So tell me, why can't we all win in this?"

I shook my head and turned away from him.

He groaned in frustration. "So much for your help." He hobbled away from me and slid himself back onto the ice. I stayed there until it was time for the ice to get cleaned up. Mindy found me, and soon after, the rest of my friends.

"Brad is such a meathead," Mindy said, rolling her eyes. "His first question was about my bra size. Can you believe it?" I nodded. I most definitely could.

"Hazel!" Evan called, advancing towards me. Fortunately enough, it was loud around the rink, and nobody was paying attention to this odd display in front of the—popular guy and evil mute girl. "I have something else to say."

I looked between my friends and then back at him. He was closer now, his breath practically mingling with mine as he cupped my face in his hands. "I'll make you help me."

Before I could escape his grasp, he let go and returned to his friends. I looked up at my friends, and judging from their faces, they were just as confused as I was. What game was he playing right now? Was this just some ploy from Sasha? Did she genuinely want Evan?

"What the hell is going on?" Ana asked, shocked. Andy's eyebrows furrowed in concentration, and then he sat down close to me. He whispered in my ear, "Do you trust me?" I was shocked by his question but slowly nodded. His warm hands grabbed my face, pressing his lips softly against mine. All the thoughts roaming around my brain stuttered and stumbled over each other. When he pulled away, he looked to make sure Grey and Evan looked.

They both noticed.

# CHAPTER 9

After we were allowed back on the ice, I skated alone for a bit, drowning in the thoughts overwhelming my mind. Why did Andy kiss me? What did Evan and Grey think? And Sasha? Faces blurred as I glided past them, and I refused to make any eye contact. I felt self-conscious as if everyone was staring intently at me.

"So, the nerds do date within the friend group," Grey said. I glanced at him, surprised by his sudden appearance. He didn't look at me, though, just skated by my side effortlessly. I came to a halt by the wall, signing, "No, I'm not dating him."

"No?" he asked, his facial expression growing grim. "So you're dating Evan? I didn't know you were brave enough to take your fill of our lot."

"What? No!" He smiled at me, making his following rude comments surprisingly disingenuous. "Sure, I get it. You've gotten braver since you got placed on the radar." I clenched my teeth together, subduing my confused, rising anger. Why was he toying with me?

"I'm not braver. You're the ones who won't stop messing with me."

"I'm not messing with you," he said.

"Yeah, well, your girlfriend is. And what you like reflects you, so I guess that's very telling then."

Our eyes locked, and we stayed there, unmoving for a bit. Then, I saw my friends coming toward us from the corner of my eye.

"You should probably go," I signed to him. "My friends don't like you."

"They're gonna pass us," Grey said, not bothering to spare them a glance. I looked at them, and they didn't look back.

"What's going on?" I asked. "What did you do to them?"

"I don't want them to talk to you right now, that's all." He held his arm out for me. "Skate with me?"

Hushed whispers built around us as I hesitantly accepted his offer. The rink cleared, and it was just him and me. Something was off and so wrong about this.

"What's your favorite color?" he asked. I glanced up at him warily and whispered, "Teal."

"Mine is auburn," he said. "Like your hair." I slipped, and he caught me, lifting me back up. He held me close to his body, and my eyes widened at our closeness. What the hell is going on here?!

"Your voice is nice," he said.

"I can't speak," I said. And then my eyes widened. I was speaking. I shoved Grey, sliding effortlessly back into the wall. I pressed my finger into my hand, and it phased through it.

I. Was. Dreaming—now lucidly.

Now things were starting to make sense. I straightened up and looked at Grey. There was no way I wanted to stay trapped in this dream with that menace. I couldn't even stand to look at him.

"Wake up!" I commanded.

He laughed, startled. "Yikes. You don't like me as much in the dream-world as in the real world." My world started spinning. How could that be possible? Unless... did he have the same talent like me? Could he go into people's dreams?

"You know you're dreaming right now?" I asked, pushing myself off the wall.

"Yes...?" he asked. But I didn't feel uncomfortable. So that means he was in my dream. He traveled into my dream?! Is that even possible?! I thought I was the only one?

"You can go into other people's dreams?"

His eyes narrowed. "Um... what?"

I shook out of my daze. I had to figure out more. With a slight smirk, I said, "Wanna see something cool?"

He nodded warily.

I closed my eyes. When I reopened them, we were on a boat in the middle of a calm sea. The sky was a vast twilight fading into a near black-purple; bright stars scattered across it like freckles on a sun-kissed face. The water and sky laced together, seeming almost infinite—it was hard to see where one ended, and the other began. Salty air flooded my nose, along with heat, even in the cool night air.

The boat we were on was a small wooden thing. Grey sat on one end, and I sat on the other. He looked around, his eyes as wide as saucers. "Wow." His reaction made me feel great. It wasn't my first time being here. Sometimes I could lucidly dream without having to be sucked into someone else's. In those rare cases, I went to the ocean. I'd lay down in the boat and look up at the stars while caressing the water, creating the only ripples on the otherwise still surface. Then, when I wanted more color, I sunk my hand in the ocean. And that was the best part.

"I know," I said, way too shy for my liking. "Want to see another cool thing?"

"Definitely." He was so strange whenever he was in dreams. It was almost as if he was a different person—he was playful. The Grey I saw at school was arrogant and a sucker for Sasha. I wondered why he changed so much into a different person whenever he was asleep?

Or was this who he indeed was? Was he a decent person over the facade he had put up?

I leaned over the boat, using my index finger to tap the water. As the ripples spread, so did a fluorescent neon blue color in the ocean; it spread to the horizon, the star's in the night sky then dulled in comparison.

He couldn't say a word. Instead, he looked back and forth between me and the ocean. Triumph flooded into my chest. I'd never done it before, but I already knew that I loved amazing him.

A bathing suit replaced my coat, not needed in this tropical air anyway. "Want to go for a swim?" His head spun toward me and took in my form. "When did you get on a bathing suit?"

"We're in a dream. You can do anything."

I stood up and dove into the water. I could see amazingly clearly; the light around made my skin glow. As I swam through the water,

I heard music all around me. The tiny creatures that lived in them serenaded me as I passed. Grey stood on the boat when I resurfaced, pulling off his shirt; he was already in swim trunks. I shyly looked away from him until he jumped into the water. He came up beside me, face glistening with droplets, hair plastered against his head.

"This is amazing," he said. "I don't ever want to wake up." For the first time, he looked delighted. He had a small smile, and his face looked brighter than usual. His eyes were calm, and they squinted as he smiled. Without a second thought, I caressed his cheek. Although he was treading water, he seemed frozen. My stomach dropped in embarrassment, and I suddenly felt ashamed. I don't know what came over me! There was no reason, and it was wrong, right? He was rude to me and also dating Sasha. It was the dream world, but he was conscious that he was there.

Gosh, I was such an idiot.

I dipped my hand back in the water and dove under the surface. I swam deeper towards the increasingly bright lights—the further away from him, the better.

Fingers clasped around my ankle, stopping me in place. Grey slowly flipped me over me to look up at him. His hands wrapped around my waist and pulled his body close to mine. His hair floated around his face and eyes. I wanted to push it out of the way, but I kept my hands where they were on his shoulders. His eyes held mine, and my heart started to flutter in my chest.

"Grey...?" I asked, bubbles floating from my mouth. He hovered above me, hands still on my waist, and I could see the boat some distance away on the surface. We sank deeper underwater together with each passing second.

"I'm going to kiss you," he whispered. My eyes widened.

"What about Sasha?" I asked. If he heard me, he ignored what I said.

Before I could get out another word, his lips captured mine as an alarm blared in the distance.

# CHAPTER 10

My eyes flew open, and I lay there for a moment, shocked. Grey kissed me. He didn't even care that I mentioned Sasha. He went ahead and kissed me.

I pushed my finger into my hand, a reality check needed after a lucid dream set in a familiar environment. My finger rested in the middle of my palm; my nail caused a slight pinching sensation. I was awake.

I hastily wrote down my dream in my dream journal next to my bed, making sure I didn't lose track of reality. I've torn through many of those, making sure I could still keep track of the real world and the dream world.

But I hated how suddenly this dream snuck up on me.

I closed the book and shook away any exciting thoughts. Who cared if Grey kissed me? Dream or no dream, he was still with Sasha, who was trying to ruin my reputation. And he was helping her. So, he didn't get the privilege of owning my feelings.

But he already did.

***

Stupid dreams. I'd do it without hesitation if I could tear my subconscious to shreds.

I couldn't even look at Grey at school the next day. I acted like he didn't affect me as I passed him; he talked with some of his guy friends by their lockers. Swallowing nervously, I averted my eyes before ours met. Grey's friends continued talking, but he went quiet, and I could feel him looking at me.

I rushed down the hallway into homeroom and to my table. Ana and Blake looked at me curiously.

"You ran in here like your butt was on fire," Blake said teasingly. Ana laughed and said, "Right? Are you avoiding a teacher?"

"No, it was nothing," I signed.

"You are such a horrible liar," Ana said in mock disgust.

I sighed and then signed, "Grey was around, so I tried to get away from him as quickly as possible."

"Did he do anything to you?" Blake asked. "Should we tell the dean or something?"

"No," I signed. "Grey did nothing. Besides, he has Sasha on his side. She'll lie about me and get away with it. I don't have the words to defend myself."

"I think you just don't want help," Blake said. "You don't have to do this on your own."

He was right—I hated help. I preferred to handle things on my own. Besides, I was terrified that Sasha would still be able to get the upper hand somehow. She was manipulative and convincing. She's had her fair share of drama in the past, and I always thought she was the victim—everybody hurt her, not the other way around. And now that I was finally the villain, a part of me felt Sasha spun every other negative interaction she had to paint her in the best light.

The bell rang, and I changed the subject as I stood. "Any tests coming up?"

"I have an Anatomy quiz," Ana said with a groan. "I can't do it."

"Yes, you can, ba—Ana," Blake said. They both looked at me, but I casually pulled out my phone.

"I have a calculus test this afternoon," Blake continued. "It should be easy since the class hasn't been too tough." I nodded. I wish I had more classes with them, but it didn't tend to work out that way because our grade is so large. So we separated in the hallway to get to the first period.

I entered ceramics class, one of the classes in school I was the happiest.

"Hey, Hazel!" said Jacob, one of my friends I wasn't close to but thought was cool. "How's it been going?"

I typed on my phone, "Good. How've you been?"

"I've been good. I have to finish this project today, though, or else Mr. Swift will kill me."

"You've been making good progress, though," I typed. I looked at Jacob's project he had placed on the table. His hands had globs of clay all over. He sculpted a baby turtle emerging from an egg, and it looked great. He messed with its nose for a second before shrugging. "It could be better. I still have so much I want to add to it. Anyway, how's your project going?"

I cringed. I wasn't nearly as creative as all my peers. I just sculpted a mug, which was why I was so close to finishing already. "It's fine," I typed.

I grabbed my project and then sat across from him. We worked quietly for some time as other students gradually filed in towards their projects. As they did, Mr. Swift told us to continue on our projects and that the deadline would be the end of our next class. Jacob squinted in concentration as he picked up the pace.

Jacob looked up at me a few minutes later like he remembered something.

"I heard Sasha was giving you problems!" he said. He looked angry as he looked me in the eyes. "Are you okay?" I smiled and nodded.

"If you ever need any help, I'm here for you, all right?" he asked. I nodded again.

At least he didn't believe her.

After Ceramics finished, I saw Grey and Sasha. They walked toward me, Sasha clutching his arm while she laughed and stared at him. Their increasing closeness filled me with anxiety, and I moved as far away from them as possible; I walked straight and avoided eye contact. Sasha's head began to turn in my direction as we got close to each other, and my stomach dropped in anticipation. But, Grey grabbed her face with one hand and gave her a quick kiss. It gave me enough time to slip past her without noticing I was there.

I was slightly relieved, disgusted, and jealous. I quickly got to Spanish class and sat in my chair with a heavy feeling in my chest.

Why did I have that dream? None of this would matter if I never entered his dreams in the first place. I would've never been on his radar at the party nor been bothered by Sasha.

This situation was all becoming crazy. I was going crazy.

***

I never tended to care that Evan was in my Spanish class. That was until he took notice of me.

My friend Nancy usually sat in the seat next to me; he took it before she got to class. I was surprised and quickly reserved the other seat on the other side of me so she could still sit beside me. Evan unpacked his stuff as he continuously glanced at me.

My teeth clenched. I kept my gaze away from Evan and strictly on the whiteboard or floor. Nancy got to class not too long after that and looked just as surprised as I felt. I just shrugged and moved my bag from the seat next to me. She sat down and then leaned over, whispering, "What's he doing up here?" I shook my head, just as lost as she was.

Did I even want to know what he wanted? Because it was something.

While the teacher was talking, Evan slyly passed me a note. I warily looked at him and back at it. I opened it slowly to see the words: I want to talk to you after class. I stole a glance at him, and he looked back. After that, I returned my focus to the lesson, although I heard nothing.

He wanted to talk to me! About what? He and I had nothing to talk about. A part of me was worried that he would deliver some bad news to me from Sasha. Was he going to threaten me with something else? What could he possibly want?

I tried desperately to get away from him quickly after class. But he pulled me to the side and trapped me between him and the hallway wall. "You and I should stick together." I looked at him, slightly terrified and slightly confused. "Date me, Hazel."

I swallowed, unable to form a coherent thought in my mind for a few seconds. My eyes searched Evan's face and his chocolate brown eyes. His hair was short and blond and covered part of his forehead. He wasn't unattractive physically, that's for sure. But, he wasn't a nice person; he cheated with Sasha.

I pulled out my phone and typed, "Why do you want to date me? You don't even know me."

"I've heard about how you've been defending yourself against those two losers, and it's impressed me. I want to get to know you more. Think about it?" I nodded because I felt obligated to, earning me a bright smile from him. He pulled me into a tight hug against his body, wrapping his arms around my lower back. My arms wrapped themselves around his neck.

It felt weird but warm. My body enjoyed the tightness of Evan's hug, the strength of his arms, and the beating of his heart.

But he's not Grey.

# Chapter 11

Later that night, I ate dinner with my mom as I weighed the pros and cons of dating Evan, not because I seriously wanted to date him, but because I was so intrigued that he'd ever want me. Even so, I couldn't think of anything good coming from the outcome of being with him. He's okay with cheating, and I knew he still wanted Sasha. So Evan wishing to date me must be a ploy, like in Grey's dream, that Evan is pulling to make Sasha, and dare I say it, Grey, jealous.

"Hazel?" I heard.

I wasn't listening to her the whole time. My head snapped up.

"Sorry," I signed.

"You all right?" she asked.

"Yeah, why?"

"You're not listening to what I'm saying."

"You want to have Jared come over and have us all spend some time together one of these nights," I signed, which was the only thing that filtered through my deaf ears.

"I guess you were listening," she said and laughed. "Yeah, I'm still thinking about it and him. I want to make sure he's a good guy before inviting him over to spend more time with us."

"Sure," I signed and had nothing else to say.

Once dinner was over, I went upstairs to finish the rest of my homework.

Stop thinking about Evan! I angrily scolded myself. My mind, though, had plans of its own and gave up on the pointless problems on the sheet.

I tried taking a break and scrolling through my Instagram feed when an unknown number interrupted a video I was watching. I opened the message, which read: hey... is this Hazel?

Me: who is this?

Unknown number: it's Grey.

Whoa. That's not who I was expecting.

Me: Oh. Hi...?

My heart couldn't help but flutter in excitement. There have been many times when I've wanted Grey's number to be housed in my phone so that I could have the ability to text him. It finally became a reality.

Grey: How are you?

I rolled my eyes at his painful small talk.

Me: Is there something you need?

Grey: I heard Evan asked you out. I don't think it'll be best for you to date him, in my personal opinion...

Me: Why's that?

Grey: Well, you saw what he did to Sasha and me. If he's okay with cheating, he'd most likely be okay doing the same thing to you.

I paused for a second, his words ringing as truthful. Reading that someone else thought the same thing as me validated my concerns.

Grey: Anyway, besides warning you about him, I have a question for you, Hazel, and I'd really like it if you answered honestly.

My heart sped up in my chest the minute my eyes soaked in those words. A question? About what?!

Me: Ok

Grey: Have any interesting dreams lately?

The questions sent a shiver down my spine. It couldn't be that Grey's talking about the dream we shared? He couldn't possibly remember that dream!

Me: Sure.

Grey: What about?

Me: None of your business.

Grey: Listen, I'm sorry we've been treating you like crap. Sasha can get a bit jealous at times, but she's a really sweet person, and I think we should all start over. We both feel pretty awful about the fact that we read that poem about your dad. I really didn't know, even if the context clues were pretty obvious.

If he wanted to start over, I would need a little more than just a few words of apology.

Me: If you're so sorry, you'll delete that Instagram post Sasha wrote lying about me and replace it with a post saying that she did it out of jealousy or anger or whatever the hell reason she did it for and admit that those texts were not genuine

He didn't respond to me for the rest of the night. I didn't know if I scared him off or if maybe it was Sasha who was texting me and not him, but it was what it was. I stopped checking my phone, achingly hoping for a response from him. It frustrated me that I felt something for him when I shouldn't have.

Wait no... I can't believe I have feelings for Grey. I didn't want to admit it to myself, but I was tired of being in denial. It was true; I felt some attachment to him. I was hoping that it mainly just had to do with the dream Grey, who was completely different than the actual Grey.

***

Sasha stopped me in the middle of the hallway at school the next day before homeroom, a fake smile on her face. "Can I speak with you real quick? I know the cat tends to have your tongue, so you'll just have to listen to me." Anger was already boiling over in my chest. I ignored her and pushed past.

"Oh, so you don't want me to clear your name!" she called after me. I stopped in my tracks and reluctantly looked over my shoulder towards her. Her phone was in her hand, and she waved it around smugly. Then, swallowing my pride, I walked back towards her.

"So, no, I'm not going to say that this was all due to jealously or anger, as my boyfriend suggested," she said, "but I will say that you apologized and that we're good friends. Deal? It's the least I could do for reading that humiliating poem out loud." She wasn't genuine at all. It disgusted me.

I knew her offer was too good to be true in the first place. So I walked away from her faster this time, even after she called after me again. I should've known Grey couldn't resist Sasha's damn charms and that he would've settled for that compromise for her.

For her.

I don't know why that made me jealous. Grey still loved Sasha. The dreams meant nothing to him. They were just dreams.

And it hurt a lot to know that.

# CHAPTER 12

Evan stopped me in the hallway right before lunch. "So, have you come to a decision?" I dreaded this moment.

It wasn't a matter of wanting to drown my feelings for Grey anymore; it was a matter of the consequences of me saying no. I slid my phone out of my back pocket and opened it up. My fingers hovered over the keyboard in my notes app, unable to process what I wanted. I repeatedly thought, I don't know Evan, he isn't above cheating, and he most likely isn't a good guy. But what else can happen to me if I say no?

"Evan!" I heard Grey call. I froze, my eyes locked on the screen. Grey approached and stood close next to me. "What's up?"

"Not much," Evan said, wrapping his arm around my shoulder and pulling me to him. "Just talking to my girlfriend." My eyes met

Grey's, his boring into mine. So, I guess this decision is decided for me. As Sasha approached us, I wanted to scream.

"What's going on?" Sasha asked. A flash of anger tore through Sasha's eyes as they landed on Evan and me. Evan noticed, and his fingers tightened around my shoulder, a smile illuminating his face.

"Just talking," Evan said nonchalantly.

"Why is your arm around Mutey like that?" she asked, and my stomach lurched in anger.

"She's my girlfriend," he said, "that's why. Do you like being called Mutey, babe?" It took me a second to realize he addressed me, and I angrily shook my head. "I didn't think so." He glared at her. "Stop calling her that."

"We'll call her whatever the hell we want," Grey said harshly. I fought to ignore the sting I felt in my chest. Sasha's eyes were bloodthirsty as she looked me up and down, saying, "Mutey here isn't even that attractive. Don't know what you're going to get from her."

"You better stop calling her that," Evan warned.

"Or what?" Grey asked, taking a step forward. I pushed my arm in between the two of them.

"Mutey," Grey said, making me want to vomit, "get out of the way." My stomach was tense due to all the conflicting emotions I was feeling, and I was fighting the urge to let them see my tears. It felt awful that the one person I liked could say such harsh things to me. I pushed Evan away from Grey and took him by the hand, leading him away from them.

"You're so sweet," Evan said. "I'm sorry you had to be around any of that." I nodded, unable to think for a moment. I wanted to cry so badly. He noticed and quickly rubbed my back. "They're just mad about us. Don't let them get to you." He walked me to my next class, English, and I practically ran toward Mindy, who was in our seats.

"Hey, darlin'," she said and then processed the awful emotions written across my face. "Oh no. What's wrong?"

I typed out as much as I could while she read over my shoulder before our teacher and Sasha and her plastic posse got to class. They all sent glares, noticed by our classmates around us. I wanted to sink further into my seat.

As the class went on, I could hardly concentrate, nor did I want to. I could only think about what I got dragged into and how Grey completely flipped the switch and turned on me. I was so stupid even to entertain the idea that he'd be nice to me.

Phones throughout the classroom and practically everywhere in the school buzzed or went off, and our teacher, annoyed, said, "Silence your devices."

I peeked at mine and saw a party invite was sent to everybody in our grade via our school emails. The sender was Evan.

Celebration time! Join me and my girlfriend at my house this weekend ;). He put his address so everyone would know where to go. My stomach dropped, and my breaths panicked. I couldn't do this. Even if I had the strength to tell Evan I didn't want to be his girlfriend, it was all gone now. Since he announced to everyone that he had a girlfriend, he'd flip if I bailed out on him now.

But I couldn't handle the fact that everybody in our grade now new. I wanted to vomit, but I don't think that would be very helpful in a moment like this.

After class, Mindy and I headed to lunch. I sulked, and she wrapped her arm around my shoulder. She didn't know what to say, and honestly, neither did I. I felt like I got myself into this mess even though I said nothing!

I arrived in the cafeteria, and as always, it was normal. Like the entire cafeteria staring at me, all my fears were never realized. I let out a small breath of relief and waved to our friends before heading to the lunch line.

After grabbing our food, we sat with our group.

"Did you guys see that email?" Ana asked. Mindy didn't share Ana's enthusiasm. "Yeah, we did."

"Who do you think his girlfriend is?" Blake asked. As if on cue, Evan wrapped his arms around my waist, causing me to jump, and planted a kiss on my cheek. "I'm gonna sit with my friends, but just know I'm very happy about us." He pulled away and looked at our table with a smile. "I hope to be seeing you all there." He walked away.

All my friends stared at me.

"Hazel, what?" Andy asked, astonished.

"He asked me to date him, and I didn't know what to do! He just decided for me, and now Grey and Sasha hate my guts again!" I frantically signed as Ana translated.

"Break up with him!" Arnold said, as if that was the easiest solution in the world.

"I can't! The popular kids already hate me, and he's not an outcast himself." I let out a frustrated sigh. "If both Grey's group and Evan's group are against me, I will have to leave this school!" Blake translated this time.

"There has to be something we can do," Mindy said, "to get you out of this position."

"Or, we let it play its course," Andy said, and we looked at him in alarm. "I mean, come on, it's obvious what he's doing. He's trying to make Sasha jealous of you two so that she'll crawl back to him. You don't have feelings for him, right?" I shake my head. "Great. So in due time, maybe even at this party, Sasha will try to get him back, break Grey's heart, and be over and done with. Nobody will even bat an eye at you. They'll be too engrossed with the new drama."

I don't want Grey's heart to break, though.

"If it isn't my favorite bastard Mutey," Sasha's voice rang behind me, and I turned to see her with her plate of food. "Looking forward to this upcoming party."

Grey said nothing. He glared at me before following her to their table.

My hot anger was at the point of exploding. I turned back to my friends, and a surprisingly calm smile painted my face as I signed, "Let's make sure Sasha destroys him."

# CHAPTER 13

After school, Evan met me outside the school with his usual smile. "You excited about our party?" No! I didn't know there could be so many parties in high school! But instead of typing what I was thinking, I returned his smile and nodded. "I knew you would be. We're going to have drinks and stuff. That's okay with you, right?" I shrugged. I couldn't care less. "Good. I'll see you later, babe." He kissed me on the cheek before walking over to his friends, who I hadn't even noticed were waiting for him a few feet away from us.

I turned to see Sasha glaring at me from in her car inside the school parking lot. I gave her a quick smile, and she flipped me off. The anger in her eyes from seeing Evan and me together sent a jolt of energy through my body. She wanted him, and she wanted me to burn.

This party was going to be very interesting.

***

The night before the event, we found each other.

I was in my dreamscape again since I felt no discomfort. I stood by a lake, the one I went to with my dad right before he got shot. Tears fell out of my eyes as I stared down at my reflection. They dropped into the water, disrupting the calm with gentle ripples, which slowly turned giant and turbulent.

My feet slowly sank past the rocky ground of the lake into the muddy sand below. The water burned my ankles as the waves I created changed course and slowly advanced toward me.

"Help!" I screamed as panic flooded my chest. "Please! Somebody!" Usually, I have control of my dreams, especially when they're lucid. But this dream seemed to have a mind of its own.

The first wave loomed over me. I closed my eyes as the water crashed over my head, yanking me out of the ground and rolling me up the rocky beach. As it receded, I gasped, pulling myself up. My clothes clung to my body like a second skin and smelled like blood.

I wobbly stood and made my way to the tree line to avoid the other wave coming towards the beach. My body felt sluggish and stung everywhere.

"Help," I weakly called. The second wave came towards the shore as I gripped onto a tree to steady myself. A hand clamped down on my wrist and pulled me up. Grey's eyes burned into mine, and he yanked me through the woods.

"What are you doing here?" I asked.

"I don't know," he said honestly. "I don't even know why I'm helping you." I could hear the rushing water behind me, edging up the shore, knocking down some of the weaker trees. It towered above us about two feet.

I tried to move faster, but my legs slowly turned into lead. I couldn't push them any further, and they stiffened.

I fell, making the wave behind me look even more daunting.

"Don't look back," he ordered, yanking me back up. "Just keep moving."

"I can't," I managed to say before the water collapsed onto us. It dragged me across the ground, more cuts forming on my skin. My body contorted, flipped, twisted, and shot through the rushing water. I couldn't get up, resurface, or even breathe. I kept my eyes squeezed shut and held my breath for as long as possible, my lungs burning.

A hand grabbed my soaking wet clothing and pulled me out of the water. I gasped for air, my back pressed against Grey's stomach as he labored for air. "I didn't realize how scary lucid nightmares could be." His voice strained as he used one arm to hold onto a tree, the water rushing around us. It was starting to smell saltier.

"Grey! Whatever happens to me, don't let go," I heard a woman's voice order.

"No," he whispered. The pine trees across from us started to form into palm trees, a woman and child attached to one of them. She had a striking resemblance to Grey, and the little boy looked like he could be Grey's brother. The woman's grip looked like it was slipping. Panic flashed in her eyes, but she covered it up with a smile to her presumed son.

"Mommy, I'm scared," he whined.

"Don't you dare let go of this tree, Grey," she ordered, flooding me with shock. "Do you hear me? Please, for me. Don't let go." He nodded. She looked like she couldn't thoroughly flush herself against the tree, her arms not strong enough to hold on.

"What about Cecelia?" he asked, his eyes wide with fear. "Won't she drown?" Her other arm rubbing his back, let go to touch her stomach. She was pregnant.

"She's okay, honey," she said after a moment. She gave Grey a weak smile. "She's still kick—" Her grip wasn't tight enough. She slipped away and let out a scream, young Grey calling after her as she was dragged off into the tumultuous waters, her head disappearing from view.

"Mommy!" young Grey screamed, frantically looking around. He wailed as he continued searching for her. A lump in my throat formed, and I knew what this was.

A memory. A dark, horrible memory. And whether or not Grey realized it, he let me see it.

Grey's arms tightened around me and the tree. "Damn it," he muttered. His younger self kept screaming and tore a hole in my heart. So this is what he held inside of him. I thought he was such an obnoxious person and didn't know what it was like to feel anything because he was rich. But he experienced loss in such a horrific manner.

"Are you okay?" I whispered.

"No," he hissed. "I'm not." Then, as the waters receded, the scene shifted until we stood next to his mother's body. She laid in an unnatural position as younger Grey found her, rubble all around them. Grey screamed as Disaster Relief workers picked up the body. He screamed as he followed after them and cried, "That's my mommy!" I didn't even realize I had started crying with him.

"They were only able to save my sister," he commented as his younger self and the workers disappeared from view. "She miraculously survived."

"I'm so sorry," I whispered. Grey let go of me and took several steps back. That's when the scent in the air began to change, and the sun started to set. Palm trees turned into trees typically seen in our town and the rubble melted into the ground, turning into pavement. Gasoline filled my nose, and I knew what was happening. I reached out for Grey, but I was in the arms of the man, who had a gun pressed into my temple. Grey stood next to my dad, his eyes wide in horror with tears from his fresh open wound caused by his memory.

Grey attempted to grab my dad, but his hand phased right through him. He looked back at me as the man snarled, "Give me the money right now, or I'll kill her."

My dad scrambled in search of his money as the man's vice-like grip got even tighter. "Faster!"

"Daddy," I said, tears escaping my eyes. Grey flinched and clenched his fists.

"It's okay, baby; I'm going to give the nice man some money, and we're gonna go," my dad said as he dug through his wallet to no avail. But unfortunately, he left his credit card in the car, which is what we were walking back to get because he had no cash.

"You have no money on you?" the man asked.

"No, I'm sorry, if we could just get to the bank or something, I'll get you the money," my dad quickly suggested. "I just have to get my card."

I closed my eyes and swallowed as the man said the dreadful words, "Not necessary." The gun went off, and my dad crumpled to the ground, droplets of blood littering my face.

I couldn't contain it like I did when I was a kid. I screamed, clutching my hair as I sank to the ground, sobs quickly following. "Dad, no, please." I crawled over to his lifeless form before Grey's arms wrapped around me and pulled me into his chest. He held me as we both cried together.

# Chapter 14

Waking up without Grey holding me caused a deep ache in my chest to burn like fire. I didn't want to admit that to myself, but there was no way of denying it. I needed him there. After what I saw... after Grey triggered my memory for him to see in return, I needed him more than anything. I sat up, my feelings about the party now destroyed. My motivation to have Sasha destroy him was my anger, but that was all gone. He was there for me. He tried to stop the memory.

I pulled myself out of bed, fresh tears leaving my eyes as I did. Dad's plea he gave the man in desperation to save my life when it should've been to spare him echoed against my skull. He didn't know he was going to die.

I resolved that I wouldn't be going to the party anymore. I was going to wallow in misery instead.

After regaining my composure, I went downstairs to the kitchen, where my mom was preparing breakfast. She turned and began to ask, "How'd you slee—" She put down the spatula she was holding and her arms wrapped around me. "Why were you crying?" Sobs escaped my chest again, and I leaned into her.

"Baby," she said and pulled away from me. "What happened?"

"I had a dream about dad," I signed. She flinched and then pulled me into her again.

"I'm sorry," she said, her voice cracking. "That must've been painful, especially because your dreams are so vivid..." I told my mom a long time ago that every dream I had was so vivid it was scary; I neglected to tell her that I entered into other people's dreams frequently because I knew she wouldn't believe me. Eventually, she informed me that my dreams were lucid, and I did further research online to figure out how to deal with them, including having a dream journal by my bedside at all times.

I pulled away from her as I sniffled, putting on a weak smile. She wiped my tears away and then turned to the stove. "Breakfast is almost ready, okay?"

I sat down at the table and waited. Mom brought the food over, and we ate together in silence. She looked at me and gave me a tight smile as if she had to break something to me that she didn't want to.

"Jared's coming over tonight." An audible hiss escaped my lips, and she looked at me knowingly. She probably made these plans a while ago, but it was awful to hear with the timing of my nightmare.

"Okay," I signed. "I'll be out of your hair."

"No... I thought maybe we should spend some time together, all of us." Nausea flooded through me, and I gripped the fork in my hand before taking a slow breath and putting it down.

"I already have plans tonight," I signed.

"Can't you cancel them?" she asked. "If it's like some party, then that's no excuse to miss this."

"This?" I signed. "What's this? I don't like him, nor will I ever. He can't replace my father."

"Nor will he ever," she said exasperated. "I just want you to try tonight. Please?"

"I'm not ready," I signed. "I'm not."

She sighed, and I knew her usual sharpness wanted to kick in. But she held it back and let out a reluctant nod. "Have fun tonight then." She stood up abruptly, put her plate in the sink, then walked out of the kitchen, Timmy following behind her. I put my head in my hands as my chest tightened, focusing on swallowing down the tears I felt.

I had to go to the party now. I didn't want to, but I had no other choice. Spending time locked up in my room while Jared was with my mom doing god knows what didn't sit well with me. Neither did spending time with them together make me feel any better.

I went upstairs and mentally prepared for the night ahead.

***

"Hey, babe!" Evan said over the music, pulling me into a hug. "How're you enjoying the party so far?" I had been there for a while, with no sign of Grey and Sasha. Evan's house was more quaint than Grey's but still spacious enough to fit people in the backyard, front yard, and upstairs rooms. So technically, Grey and Sasha could be anywhere at the party. But with the mass of shouting teens jumping and grinding to the music, it was almost like finding two needles in a haystack.

I pulled out my phone and wrote, "It's a bit loud." He chuckled and teased, "Yeah, that tends to be the case at parties." Another one of his friends pulled him into a drunken hug, and they started to stumble around for a moment. I smiled slightly. The guy walked away, and Evan pulled me close to him and said in my ear, "This party is for you. I felt you deserved better than the last party you went to. That just shows I'm a better host than Grey." He winked. "Come, dance with me." I shyly shook my head.

"Come on, no one's looking," he said and gestured around. He wasn't wrong. I'm sure everyone was too drunk even to realize I was there, nor care about it. "I want to show off my girlfriend." He dragged me to him, and we danced to the music. He was more into it than me; I held back slightly. He wasn't having that, though. He grabbed my arms and began leading the dance. I went with it, laughing slightly. I'd be lying if I said I wasn't enjoying myself.

We danced together for a few songs, him cracking jokes ever so often. We danced for a couple more songs, and then I wrote on my phone, "I need to go check up on my friends. We can catch up later, though." He nodded and then placed a kiss on my sweaty forehead. "I'll be with one of my friends in the kitchen. Shoot me a text or find me if you need anything." I nodded and then went back over to my friends, who were pressed against the sliding glass door leading to the backyard pool due to the overwhelming amount of people. It was too chilly for anyone to venture into it, though.

"Have you seen Grey or Sasha yet?" Arnold yelled nervously. I shook my head and shrugged. Maybe they wouldn't end up coming? The thought of that made my stomach churn. Then I'd be trapped with Evan for longer than I'd like. Not that I had anything against Evan.

I just knew he was using me, and so, in turn, I was going to use him, too.

After a few more agonizing minutes of drinks and loud music, I spotted Grey in the crowd, entering the house with Sasha on his arm. He looked bored already, scanning around the room for someone. I looked away before his eyes roamed in our direction.

"He's here," Andy said. "Time to get them away from each other."

"How?" Mindy asked.

"You'll see," Ana said, a smirk on her face. She pushed her way through the crowd and stood right in front of Grey. Blake smiled knowingly. Grey's eyes flickered with uncertainty as he looked at Sasha. However, her smile seemed genuine as she unhooked her arm and walked off, nodding in approval.

Andy turned to Blake and said, "You remember the plan, right?"

"Yep," he said, pushing his way past the mass of students and Ana, who was animatedly talking to Grey, who seemed to look at her appreciatively instead of cockily like I thought he would. What exactly was going on here? They refused to involve me in the plan so that I wouldn't be stressed out. But now, not knowing was causing me anxiety.

"Hazel, once Blake gets Sasha and Evan in the same area, he's gonna get you and then Grey," he said. "If all goes according to plan, that is."

After a few more minutes of agonizing anticipation, Blake came running back toward us. "Come, Hazel. There's something you should see." He then mouthed, "Act concerned." I obeyed, furrowing my eyebrows and running after him. I felt Grey's eyes on us as we passed by him.

Blake had me tiptoe up the stairs and down the hall towards a less populated house area. There was a door there, already closed. He pointed at it and whispered, "Cup your ear against it." So I did and heard heavy breathing and occasional kisses.

"Evan," Sasha breathed. "After this, you have to break up with that freak, Hazel. That's the only way I'll allow this to continue."

"Sure, baby," he whispered before planting another kiss somewhere on her body. "Anything for you." She giggled giddily, thinking she'd won. If I wanted Evan, I'd open the door, choke him, and then her. But she was getting what we wanted. She was also about to get what she deserved. And so was Grey. He would regret ever calling me Mutey.

I stood outside the door, occasionally listening to make sure they continued their session while Blake left to fetch Grey. When they came back, Grey looked confused to see me there. Then his eyes darkened.

He immediately rushed to the door and listened in like I was.

"Mother fu—" He threw it open.

# CHAPTER 15

Sasha and Evan scrambled away from each other; Sasha yanked the sheets over her naked body, and Evan fell off the bed while throwing on his pants.

"Oh no, Hazel, I didn't mean for you to see this," Evan said, sounding genuinely sorry while advancing toward me. He was a damn good actor.

Grey stood between us. He put his arm out in front of me, pushing me behind the rest of his body.

"You're a piece of crap," Grey said harshly. Evan ignored him and tried to push past to get to me. I took a step back and avoided his eyes.

"Stay away from her," Grey said, shoving Evan back.

"Get away from me!" Evan shouted. "She's my girlfriend."

"You think that she's still going to stay with you?" Grey asked. "Is that what you both think? You two wanted each other. So have each other." Grey shoved Evan back again, then led me out the door, slamming it shut. I stood still as he angrily clenched his fist and then turned and punched the door. The wood cracked beneath his weight.

He regained composure and walked down the stairs without another word. I followed after him quickly, pushing past the mobs of drunk teenagers. When I finally escaped into the night air, I saw him angrily kick in Evan's car door. I ran up to him and stood between him and the car.

He looked enraged before saying, "Get out of my way, Mutey!"

"You're gonna do more harm than good!" I signed.

"She cheated on me!" he exclaimed, running his hands throughout his hair, pulling at it. "Again! And with the same person! He was supposed to be my friend!"

"You don't need them!" I signed. "They aren't good friends. Don't stoop to their level."

His voice strained as he responded, "Get out of my way."

"No." I stared him straight in the eyes as he stepped toward me. "I won't ask again," he hissed. I placed my hand on his chest as he took another step. His heart spasmed under my palm, pulsating so hard it could burst out of his ribcage. He flinched and took a step back, swallowing.

"Why won't you let me do this?" he asked. His voice sounded too vulnerable, his eyes unable to meet mine as I stepped closer to him. He didn't move. I hesitantly wrapped my arms around him, encasing him in a hug.

For a moment, he was still—his heart was the only thing that seemed to still stir in him. But then, he returned; he gently pried himself free of my grasp. "I don't need your pity, remember?"

"I don't pity you. I... I know you're hurting."

He chuckled bitterly and rolled his eyes. "You know nothing about me. I'm not even hurting. I'm just angry."

"I know a lot more than you realize." Hazel, you're giving away too much. Bite your damn tongue-fingers.

"We've hardly even talked!" he said exasperatedly.

He stomped away from me and went to the other side of the car to begin vandalizing it again. I shoved him back and signed, "This won't help!"

"Says who?" he shouted. "He'll have to deal with getting a new car, and I get the satisfaction of knowing that. Move!"

"Why does this have you so angry?"

"It's not just this, Hazel. It's everything. I'm angry at everything that's happened in my life. You just don't get it. You don't know what I've been through."

"Yes, I do."

"You know nothing! Stop trying to play mediator. You're just making me angrier. You think you can understand because you're compassionate or whatever you think you are, Miss Goody Two-Shoes, but I know you're nothing but a mute girl who's angry. So shut the hell up, damn it!" He kicked the car next to me, and I yelped slightly. "Just shut up!"

"Grey, please..."

"I'm so fed up with this damn world. Sasha was the only one who helped me. She understood what I went through. Evan, too. And what did they do?" He laughed bitterly. "And now you want to help? Why? You're just like them." He added slightly under his breath, "Everyone is fake. I'll never let people know about my trauma ever again."

"I already know your mother drowned... I get the pain of losing—"

"How the hell do you know my mother drowned?" he asked coldly.

I didn't even realize I slipped. I froze, and Grey stared at me, waiting for an answer. We stayed like that for a moment. My eyes glued to the floor in panic.

"How... do you know that?" he asked. My eyes remained staring down. "Hazel." His voice was laced with anger. "How do you know that? Did Evan tell you?" I shook my head. "Liar."

"He didn't tell me," I finally signed, my eyes meeting his.

"There's no other way you'd know that!" he shouted. "What, do you stalk me? Is that how you know? Mutey, the Stalker, is that it?"

"No!"

"Then tell me how you know," he growled, forcing my back into the car. I shook my head. "Tell me." His hands encased my face, causing my breath to hitch in my throat. "Who told you?"

"What are you doing?" Andy shouted. A relieved sigh escaped my throat as Grey released my face and took a step away from me. I steadied myself as Andy came to my side. Ana, Blake, Mindy, and Arnold followed suit.

"You okay?" Ana asked me. I nodded and leaned into her as she wrapped her arm around me. Grey didn't spare them a second glance. Instead, he walked away from us, his hands in his pockets, almost as if he was never angry. But, for some reason, he let me know he was.

"What was he doing to you?" Andy asked. I shook my head and waved the question away.

"Well, our plan worked," Blake said. "Good job, everybody."

"I feel so bad for him," Ana said. "When I gave him ideas about dates he could have with Sasha, he seemed genuinely happy to hear about them. Big oof."

"If anything, he'll get back together with her," Blake said. "But we've freed Hazel from Evan, and that's all that matters."

"Are you sure Grey will get back together with her?" Mindy asked, causing us to all look at her. "Like... what was going on over here, Hazel?"

I pulled out my phone and wrote, "He was angry, so he took it out on me."

"That's not my definition of taking it out," Mindy said. "It seemed like he was full-on, ready to kiss you passionately." I felt my cheeks heat up, but I shook my head and hurriedly typed, "Trust me, he wouldn't have. He was being intense."

"Sure," Mindy said with a smirk teasing her lips. I turned away before she was able to read too much. In all honesty, I didn't know what Grey wanted to do. He was so close to me.

The thought of his lips touching mine sent a flood of butterflies swarming into my stomach. My steady breathing broke for a second, and I hoped nobody noticed.

"Well, now that this party has officially crashed, you guys want to go home?" Ana asked.

"Nah, let's hang out somewhere else," Andy said. "There's a nice lake around here we can go to."

My stomach dropped.

# Chapter 16

I stand at the lake's shore, my father's arm around my shoulders.

"How you feelin', cupcake?" he asks. I smile and look out at the vast expanse of the water before me. The calmness of it all soothes the anxiety I've gotten from my nightmares; every night, a man had entered a random woman's house and killed her in front of me. When he realized I was there, he'd smile and say, "Nice to see you again, my little accomplice." It was becoming unbearable.

My dad noticed my unhappiness, so he took me out to our favorite summer spot: the lake.

"I feel better," I say.

"Not worried about that nightmare anymore?" he asks. I shake my head vigorously.

"Good," he says, kissing the top of my head. "Daddy will always be with you; you know that?" His hand engulfs mine, and we stand together, listening.

A singular tear fell on my face as I stood at the lake's edge. My dad's warm hand was no longer there, and the sunlight faded away.

"You all right?" Mindy walked up next to me and nudged me lightly. I nodded and hastily swiped the tear off my cheek. Her eyes roved about my face, but mine remained locked on the lake.

"Let's walk around," Andy suggested. I stood rooted in place. "Don't you wanna come, Hazel?"

"I need five minutes," I signed, and Blake translated.

"All right... we won't be far," Andy said. "You see that rocky ledge over there?" I nodded. "We want to sit there and hang for a few. Meet us, okay?" I nodded again.

When their footsteps receded, more replaced them. I looked up, and my eyes met Grey's form beside me. "What're you doing here?"

"I followed you guys," he said and stared me dead in my eyes. "I still have more to say."

"Yet you called me the stalker," I signed angrily. "You stalked us here and waited for my friends to leave. Don't you think that's creepy, jerk?"

He ignored me and stated, "This lake means something to you." I was surprised he was so bold.

"What makes you think so?" I signed.

"I saw it in my dream. You were here screaming for help. You couldn't move, and a tsunami came in—"

"Don't you hear yourself? It was just a dream." I looked back out at the lake and clenched my hands into fists. A fog rose from the clear water, mixing with the cool night air.

"In that dream, I relived my mother's death again. And then I experienced your father's."

I swallowed and then signed, "You have one crazy imagination."

"He was wearing a dark gray t-shirt with Crunch gym's logo and gray shorts. His eyes were a deep blue, and he had dark, short curly hair. Am I right?"

"It was just a dream, Grey—"

"The robber was wearing a mask and dark sunglasses. He spoke with a lousy-sounding British accent. His clothes were all black. He had—"

I frantically covered my ears, unable to hear anymore. I tried hard to have self-control, but Grey's words stabbed deep into me and twisted.

"Sorry... if this wasn't a real memory, why would you be acting this way?" he asked softly.

"It was just a dream," was all I could sign.

"It can't be. If you're the Hazel I've spent so much time with, in my dreams, then you're why Sasha cheating on me hurts my pride more than my feelings."

"I've got to go." I attempted to take a step away, but he grabbed my wrist firmly, not so hard that it hurt me.

We stand in a tense silence. I heard my friends laughing together as they stepped out onto the rock. He took two long steps closer to me and enveloped my hands into his. The heat of his hands sent a flood of warmth throughout my body. "You've been in my head, haven't you?" I shook my head, unable to use my hands.

He sighed frustratedly. "Hazel, I've got to know how you know my mother drowned. Because if it isn't the dream, then Evan told you. Did he tell you?" I paused and looked into his eyes. Those vulnerable—almost sweet-looking—eyes. There was hope, but I had to break it. I removed my hands from his.

"I've got to go."

"Hazel, this isn't fair. You know something very personal about me, and you're gonna walk away without telling me how you know?"

"I can't tell you," I signed. "You'll have to accept that. Just like I had to accept all you and Sasha did to me."

"I didn't know what you went through. I didn't even know how drastic it was until our dream—"

"Your dream."

"I know you were there!"

"You're insane," I signed, taking a step away from him. He rolled his eyes. "Normally, I'd have to agree. But you refuse to say how you know when it would be so easy to say, Evan."

"Maybe I'm afraid you'll hurt him if I say he did."

"Well then, I promise you I won't lay a hand on him if that's what you're so worried about. Did he tell you?"

"Don't worry about it."

"You are so annoying—"

My anger boiled over. "Me?! You're the obnoxious idiot! You let everyone manipulate you, and then you come over here trying to do the same to me! Of course, Sasha cheating on you hurt your stupid feelings! You went to go smash Evan's car in because of it!"

"That doesn't mean—!"

"I don't want to hear your excuses," I continued signing, "Leave me alone! Stay out of my life."

"That's not fair, Hazel; I never was trying to mani—"

"Manipulate me? Lead me on?" Stop ranting. "To have me believe that you cared about me? Then you kissed me only to treat me like garbage? And then the icing on the cake, leading me to believe we could be friends until I didn't follow your advice regarding Evan?"

He was silent before slowly saying, "We never kissed, Hazel." I processed the words he told me and replayed them several times. The realization that I mentioned another dream struck me hard in the stomach.

I met his eyes which were unreadable.

"I kissed you in my dream, and I felt like I was cheating on Sasha." His gaze held me in place. "If our dreams aren't connected, how did you remember that kiss?" I felt a lump in my throat. He knew. "We went ice skating and swimming, right? And I dove after you and kissed even after you reminded me about Sasha?" I tried to make my eyes just as unreadable as his. My phone began to buzz repeatedly in my back pocket. I pulled it out to see Mindy calling me. Before I could pick up, he snatched the phone out of my hand and put it in his pocket.

"Give me—!"

"Hazel." Grey's voice was firm. "Is that when I kissed you?"

"Give me back my phone, Grey." I reached for it, but he grabbed my wrists and pulled me into him.

# CHAPTER 17

We were way too close to each other. My wrist was sandwiched firmly between our chests, and I hardly had the strength to look up at him, let alone stand up straight. My legs felt like jello, and my heart thundered in my chest with anticipation and sickening nervousness.

His arms wrapped around my waist and the nerves jittering around exploded in excitement. Too many sensations cluttered my mind. I stared at a single spot on his chest. I wanted to get lost in it all—his arms were firm around me, the surprisingly pleasant smell emanating from his breath, and we were nearly flushed against each other. It all felt so right.

I don't think he anticipated us being in this position. He didn't move for a moment; he was just as still as I was.

His hand slowly moved up my back and reached around to grab hold of my chin. He looked me in the eyes.

"Can I kiss you...?" His grip on my face loosened, probably so I could nod or shake my head. I didn't know what to do.

"Get away from her!" Ana exclaimed. I spun around as I yanked myself out of his arms. Mindy stood there with Ana and Andy. They must've come looking for me when I didn't pick up the phone.

Andy stepped between Grey and us. Grey looked back at me, almost desperately, and I looked away.

"Her phone," Grey said through grit teeth, and Andy reached for it, snatching it away. Then, after a few moments, Andy hissed, "Good riddance." I looked up and past him. Grey was gone.

"Let's get back to Blake and Arnold," Andy said, grabbing my wrist and leading us down the path.

***

We hung out for a while longer before we decided to call it a night. Mindy was silent for a bit on the drive home, which was unusual. I glanced at her several times and shifted uncomfortably. Once we were finally outside my house, she parked the car and said, "So." I pulled out my phone, ready for whatever was about to come.

"I like Andy."

I blinked. Mindy's confession was confusing. They always seemed to butt heads, even back in middle school. So the fact that she liked him was shocking—in a good way.

She laughed and said, "Yeah, I know, I know. I like Andy. And you're probably wondering what that has to do with tonight." I nodded. She took a deep breath and whispered, "I'm positively sure he likes you."

Another blindside and another blink. "You didn't know?" I shook my head with a frown.

"But, you like Grey, right?" she asked hopefully. "'Cause, the last thing I want is to like the same person as a friend I hold closely in my heart. Because no matter what, I can't help it; I'd have to let you have him. And that would hurt so much. I wouldn't be able to stand it and—"

I waved my hands and pulled her into a tight hug. She sighed. I pulled away and then typed, I don't know how I feel about Grey, but I definitely don't like Andy.

She was back to normal as she exasperatedly asked, "How can you not know how you feel about him? You don't let any guy hold you like that."

Frustrated, I typed, I'm sure he'll be back with Sasha tomorrow morning. I can't waste my time viewing him any other way.

"I can't argue that," Mindy said, much to my disappointment. "But you both have feelings for each other. I don't know what it is with you guys, but I feel your bodies fit together perfectly."

I laughed and typed, Our bodies? Not even our personalities?

"Hey! You think I'd believe that one of the brightest, bravest, funniest, and loyalist people in this school would honestly fit in well with one of the notorious bad boys who's supposedly also a softy?" She gave me a giant eye roll. "He's a good actor, that's for sure, but I know you could do so much better. But at the same time, you're both cute together. So..." She shrugged and let out a giggle.

I smiled slightly.

"Well, I'll see you Monday, sweaty," Mindy said, and I laughed. "Yeah, you're not a sweetie. You're a sweaty." We hugged, and I got out of her car. Right as I closed her door, Jared walked out of my house. We locked eyes, and I felt sick.

"Hey, Hazel!" he said with a big smile. I gave him a strained smile in return. "I was just leaving. But we should have a game night soon, you, me, and your mom. Yeah?" I nodded hesitantly.

Mindy was still sitting in her car, watching the scene between the two of us. I knocked on her window and signaled for her to leave. She complied, driving off. As she did, he gave me a one-armed hug and pulled away, still wearing a smile. "It's getting late, kiddo. You should go inside. But I'll be seeing you again soon!" Unfortunately. He headed off to his car, and I quickly got inside my house, slamming the door shut. I hate him. He will never, ever, EVER—

"How was the party?" my mom asked. I spun around, startled. She sat on the couch, TV remote in hand, looking deliriously happy. He makes her so happy.

"It was fun," I signed.

"That's good," she said. "Want to watch some TV?" I nodded and sat down next to her. She wrapped her arm around me, and I leaned my head onto her shoulder.

***

Discomfort spread through my chest, and I knew I was in somebody's dream. I had a feeling their name started with a G and ended with a rey.

I walked down a cobblestoned dark alleyway. The buildings towered above and beside me, and the air smelled like sulfur.

"Hey." I turned and looked up at the figure behind me. Grey's intense stare met my gaze.

"It's not safe to be here this late," he said.

I shrugged and said, "I can handle myself."

"Even if it worries me?" he asked.

"How about you stay worried about Sasha?"

The minute I mentioned her name, a figure appeared at the edge of the alleyway, draped in the street lamp yellow light. It was slender with one hand on its hip and long straight hair falling past its shoulders. Great. She's here. She walked towards us seductively.

"Sasha can do whatever she wants with her life," Grey grumbled. I rolled my eyes and walked up to her. She smiled cockily and wrapped her arms around my shoulder, forcing me to face him.

"Grey, baby," she said, "tell her who you truly want. She won't mind." Grey's eyes hardened.

"I want Hazel."

Her grip loosened, and with a broken voice, she whispered, "What?" His demeanor changed. He took a step back, his face sad. "Sasha, I'm sorry. But I can't help how I feel about her. And you've not once-respected my feelings for you. You always went to Evan..."

She broke down behind me, and I tried my hardest not to roll my eyes. Her claws dragged down my back as she sank to the ground, and her sobs echoed around the alleyway. I looked down at her, and she grabbed my hand. She said, "Please tell Grey to get back with me with tears in her eyes."

"Sasha... I've always been there for you when you cried," Grey said, all the feeling in his voice gone. I was slightly startled. "You always went back to Evan. Always. If he's so much better than me, just be with him. Go."

Her fingers released their grip, and she stood up. She wiped the tears off her cheeks and smiled at Grey. "I never wanted you anyway." Then, she walked off, leaving Grey and me behind.

"I meant what I said, Hazel," Grey said. He almost looked shy as he said it. "I can't explain why I like you. But I've dreamt about you so much; I feel like I can't get you out of my system." He wasted no time shortening the distance between us. "And if I can't get you out of my mind, why fight it? Why not get the girl I want?" His fingers tangled in my hair as they brought my face closer to his lips. "Is it the same way for you? Can you not get me off your mind?"

He searched between my eyes. Our breaths mingled and heat flooded into my cheeks. He was so close to me. I know I was close like this in real life, but it didn't make it any less real feeling in a dream. A part of me wanted to look away because of how shy I was, but his hands held me in place. He was teasing me with this closeness—just kiss me already!

I exhaled shakily and whispered, "Yes."

That was all he needed to hear. His lips crashed down on mine. They danced with each other as he pulled me closer to him until there was absolutely no space between us. My arms wrapped around his neck as his snaked down around my waist. Our lips broke from each other for a moment before colliding again. My whole body felt like it was on fire. My lips tingled, and I felt like they weren't getting enough attention from his. I never wanted to stop.

He helped ruin your reputation.

I pulled away, slightly breathless. Grey looked flustered. "What's wrong?"

"I can't do this," I said, trying to squirm away. He was reluctant to let go. "You ruined my reputation in front of the entire school, and you expect me to just... be your girlfriend? Your hookup? What do you even want from me?"

"I want to date you, Hazel," he said, slightly offended.

"Well, regardless!" I said. "You hurt me and now want me? How about making things right? Why would I just get with you?"

"I will make this right," he said thoughtfully. I shook my head and said, "I don't believe you."

"I will. And when I do, I will clear your name of everything that Sasha and I were a-holes for. And then, you and I will date. Okay?"

"Why should I date you after all you've done to me?"

Before he could reply, my mom said, "Hazel, honey, wake up." Her voice flooded the alley. "Come on, go upstairs."

Grey looked around and then back at me with his eyes wide. "I hear her."

# Chapter 18

My mom shook me awake before I could say anything else to Grey. "Come on, get to bed. We fell asleep." I was more awake than she'd ever know. The dream was still burning fresh in my mind, and I quickly engraved as much of it as I could into my dream journal after I stumbled up the stairs into my room. Will he even remember? Will he recognize my mom's voice?

Would it even matter if he did? I quickly succumbed to the one person who was trying to play me. He was scheming against me with Sasha and Evan, for all I know. He wanted something more sinister than a mere innocent relationship with me. And besides that, I doubt he'd remember any of this, especially hearing my mom's voice. He didn't seem like he was in a lucid state. And I was definitely in his dream. He wasn't in mine. He probably didn't even have the same ability as me. I, for whatever reason, didn't realize I was in someone else's dream for once—probably because I've gotten so used to being in Grey's.

I was trying not to like him any more than I already had. And when it came down to it, how far would it go? I wasn't interested in a relationship just for it to be all fun and games until someone went to a different college, and then we ended up in a long-distance relationship and—

Once again, I was getting ahead of myself. Grey and I were, not in a million years, getting together. And that was that.

The haze of sleep slowly overcame me, and I finally drifted off to sleep.

"I'm guilty."

Discomfort spread through my chest, and a familiar parking lot appeared around me. A gun was pressed to my temple, and I saw my father falling in slow motion in front of me. Sickening dread coursed through my veins, and my mind screamed for me to wake up.

"I didn't want to kill him." His voice sounded altered as it echoed throughout the parking lot. Tears flooded my vision.

"I just... I didn't realize what I had done until it was too late."

The man's grip on my shoulders tightened. But, again, I was a small child; his hands made me feel trapped underneath their hold.

"It was a split-second reaction... I just lost my wife. I had two kids to feed if I had just waited one more day—you idiot. You absolute idiot." The man slammed the gun down on the ground, and his ironlike grip released me from its grasp. I yanked myself away and fell over my dad's body, gripping him tightly.

"I'm so sorry..."

It can't be my dad's murderer.

"Honey... you haunt my dreams every night. I need peace. Please."

My cries intensified, and I cradled my father. His head lolled back as I lifted him. My arms shook as I placed my head on his shoulder, avoiding his blood.

"I didn't mean any of it," he whispered. He sounded so guilty. His hand met my shoulder and squeezed it softly. "If you could only forgive me... I beg of you."

I lifted my head as I screamed, "Wake up!"

***

I could hardly function in school on Monday—my dream left me feeling sick to the stomach. No matter how much poking around all my friends did, I remained distant. They wouldn't be able to understand no matter how hard they tried. What could I tell them anyway? Hey, I can travel into people's dreams, and I so happened to travel into the dream of my father's killer, hooray! This dilemma was my problem and my problem alone.

The cherry on top of my bad day was seeing Grey after school. He leaned against my car, staring at me expectantly.

As much as I wanted Grey, I knew he was nothing more than the trouble I'd need to avoid. Sure, dreams with him were fun. And that's all it should be. Dreams.

I signed, "Please, move."

He ignored my request and asked, "Can we go for a walk?" His voice was way too gentle, and a blush rose to my cheeks even through my unhappiness.

"A walk?"

"Yeah. Through town. If you want, we can meet there."

"I... I'm having an awful day. I want to go home."

He frowned. "I'm sorry about that. Can I try to make it better?"

The murderer's voice echoed throughout my brain. "Honey... you haunt my dreams every night. I need peace. Please." I closed my eyes as the tears threatened to spill, brimming over and pushing their way through the shut lids.

"Hazel?"

My hands trembled as I signed, "Move!" His eyes went wide as they met mine.

"Are you—" I shoved him away from my car door and threw myself inside.

I peeled out of there before he could say another word.

***

Mom was still at work when I got home. I threw myself onto the floor in the living room and began to sob. Timmy whined as he tried to lick my face. He scratched at my arms as I lay face down on the carpet. I lay there for a while as images of my dad swarmed my mind. Finally, they all circled back to the one of his body lying there on the floor, bleeding out. I will never get that image out of my mind.

I'll never be able to forgive the man who took him from me.

He got away with murder. He took my father with little remorse. He didn't stay, and he didn't turn himself in. So how could he feel bad? He didn't care when I was standing there in his arms. He didn't care when he pulled the trigger. He didn't care when my dad fell to the floor lifeless. He didn't care that he murdered my father in front of me when I still thought the world was a beautiful place. He didn't care about the trauma it would cause me. He only cared that he wasn't getting any money from my dad.

Dad. I miss you so much.

After some time, I managed to pull myself together before Mom got home. I stood up from the floor, much to Timmy's relief. He started jumping around me as I walked to the stairs and then ran up them. He waited for me at the top of the landing, his tail wagging wildly. I pet him as I walked passed him into my room and bathroom. I washed my face quickly and applied a cool cloth to soothe any swellings.

By the time Mom got home, it was as if nothing had ever happened. I could be happy for her as always.

***

No discomfort, but lots of pink that night in my dream. The pink blends with hues of blue and pastel orange. I stood amid a wonderland. The trees' leaves were a dark pink, the grass was a vibrant shade of blue, and the sky was the first thing I saw—with pinks, oranges, and reds mixing seamlessly together.

"Do you like it?" Grey asked. He stood before me between it all, a large smile decorating his face.

"You did this?"

"Of course. For you."

My heart warmed. I looked down with a shy smile. "Why?"

"You were upset," he said. "I wanted to cheer you up."

"Well, thanks," I said. "I guess."

He smiled and took my hand. "You up for swimming?" I nodded. He led me toward a lake and then swept me off my feet. I yelped, surprised, and he gave me a mischievous smile. He tossed me forward, and I quickly dreamt up a bathing suit before splashing down onto the colorful surface.

I opened my eyes. It was like a bath bomb surrounded me. The water was pink and gold, and glitter finished the look. Fish swam by, their scales glistening a rainbow-chromatic color. I reached out and touched one. It quickly slipped through my fingers and gracefully went on its way.

I resurfaced to see Grey sitting by the shore. He dipped his feet.

"It's so pretty," I said.

"You're so pretty." A content smile tugged at his lips as he looked at me.

Emboldened by his words, I said, "Come in the water."

He smiled. "As you wish." He slid in underneath the surface, and when he came back up, glitter drenched his whole body.

I broke into giggles, and he playfully glared at me. His arms wrapped around my waist and his nose pressed against mine.

"Hi," I whispered. Grey answered by pressing his lips against mine softly. I wrapped my arms around his neck as his arms tightened around my waist. A part of me knew I shouldn't have enjoyed it. But I couldn't stay in denial any more. I liked him—a lot.

He pulled away and placed a kiss on my forehead. When I examined his face, his eyes looked off into the distance and hardened.

"Are you okay?" I asked.

"More than okay," he said firmly. He looked back at me. "Please, wait for me. I have so much crap to fix and so much to prove to you. I owe you so much more than I'll be able to give you."

"Okay."

We spent the rest of the dream talking about stuff that I couldn't imagine us staying outside a dream. Everything we felt, our fears, hopes, and even how we viewed each other. It was the closest I've ever thought to a person. And I didn't expect it to be Grey.

He had me wrapped around his finger.

# CHAPTER 19

Grey avoided me like the plague until the end of the school year.

It hurt so much. At first, I thought he was trying to "make things right," but he didn't do anything. My reputation was still tainted; Sasha remained at the same table with him, Evan rejoined them, and school pushed on typically. There were no more wild parties, much less drama than before—it almost felt like the calm before the storm. Things were too peaceful.

There were moments when I thought Grey was going to talk to me. We made eye contact several times, but he was quick to look away whenever we did. And he actively avoided looking at me or in my direction. I don't know what happened after the dream, but it was as if it never even happened. And that's what hurt the most—we shared so much personal stuff, only for him to distance himself from me.

I felt worthless because of him without having a valid reason. It was a dream. Who's to say he even remembered it?

So, as I said, school pushed on as usual with the addition of me having a broken heart and no one to tell. I was miserable but smiled through it. I acted like I was okay. For the most part, I was. since Grey let me see his true colors before we advanced in whatever was blooming between us.

As summer break was approaching, my heart was finally starting to heal. Then came the end of the year party. Everybody was invited. And can you guess who was throwing the party?

The one and only Grey.

For once, my friends, and by friends, I mean Mindy, didn't get excit-ed.

"Whenever we go, it causes drama," she said as our group hung out in one of the middle school parks. Whenever the kids are away for the summer, and the park is open to the public, we'd always go there. It became our usual hangout/meeting spot.

"Yeah, and knowing Grey, he's already back in Sasha's pants," Blake said, disgusted. "He's a tool." So, we had a pretty firm resounding agreement that night's party was not for us.

That was until Grey messaged me when I was home after the hang-out.

you comin tonite?

His text message was the first time he had acknowledged me in months. Even in our dreams together, he ignored me. He walked out of my life and now was trying to walk right back in. My heart slightly skipped, but I regained composure immediately and responded, No.

I promise you it'll be worth it

No.

Ok, suit yourself

I closed my screen and tossed my phone to the other side of my bed. Timmy looked up at me with doe-eyes.

"I'm not going, Timmy... you can't make me," I signed. He looked at me in utter confusion and then laid his head down with a heavy sigh. I let out a sigh of my own, and I found myself creeping towards my closet, picking out a cute outfit. After I was satisfied with what I was wearing, I messaged the group chat and let them know I would be going to the party.

AnaR u crazy?! Y?! I thought we all didn't wanna go

BlakeY do u wanna go? Y yyyyyyyyyyy??

AndyHAZEL

MindyI'm down

AnaNo!! Don't encourage her!!

BlakeI can't believe you've done this

AndyHonestly jokes aside, don't think this is a good idea

Blake

ArnoldIf y'all don't stop spamming my phone I swearrrrrrr I'll end you ALL

BlakeLook at the chat!! Hazel wants to go to the party!

MeGuys relax... you don't have to join me. I'll be fine.

MindyLike hell, I wouldn't join you. I'll be picking you up once I finish getting ready

AnaFineeeee. I'm comin 2

BlakeWe've got ur foolish back sweaty

AndyI'll b there

ArnoldI'm staying home on this one. Have fun peeps

MindyWe shall bc you won't be there

ArnoldThat hurts...

Mindy

Mindy picked me up an hour later.

When we arrived at the house, the typical scene of teens partying with recent graduates and us newly seniors was in full swing. Grey was sitting outside his home on the grass. Several girls surrounded him, batting their lashes. One tried sitting on his lap, but he gently pushed them away with a charming smile.

His eyes met mine. "I'll see you guys later." He got up and walked towards me as the girls watched, confused. Mindy gave me a knowing glance and then took several steps back towards her car. Grey watched her go and then looked back at me with a small smile.

"I thought you weren't coming."

"'Guess I changed my mind," I signed nervously.

He slightly smiled. I wanted him to hold me somehow—to close the distance between us. But this wasn't a dream. He wouldn't dare touch me. "I'm glad you're here."

"Should I be flattered?"

"Yes." He smiled again. "Enjoy the party." He walked off into his house, and Mindy returned to my side. "Do I even want to know what's going on?"

I pulled out my phone and wrote, I like him and have no idea wth he feels about me.

A smile tugged at her lips. "I think Grey likes you. He's figuring out how to show it." Hope spread its wings in my heart and soared. I had continued denying how I felt, especially with him avoiding me. But maybe things would lookup.

Maybe.

Sasha rounded the corner of the house with Evan just as Ana and Blake rolled up to the house. She smiled tauntingly and planted a firm kiss on his lips. I blankly stared back at her and then smiled. She narrowed her eyes as she sauntered over to me.

"What's up, Mutey?" she asked. I shrugged. "Cat's got your tongue?" I nodded and tried to keep my face as void of emotion as possible. Her face contorted in rage as she hissed, "Disabled garbage."

"You're disgusting," Mindy said. Sasha gave her a sickeningly sweet smile. "Shut up, fattie." Mindy's eyes widened as she looked down at her body. She was not even close to being overweight, but I could tell that comment struck a nerve with her.

"Sasha!" Evan snapped. She glared at him while he led her off, and he shot us an apologetic look.

"I hate her," Mindy whispered. Those were my thoughts exactly.

I responded on my phone, You're not fat. She's a piece of garbage and is taking it out on everyone else. Mindy nodded, but she didn't look convinced. My blood began to boil. Coming for me was one thing, but attacking my friends was another story. As I built up my courage to look for Sasha and do God knows what, the music stopped in the house.

"Hey guys, I have a quick announcement to make," Grey's voice echoed out to the yard. Blake and Ana, who had just pulled up to the house in Blake's car, rushed over to us, visibly worried. "You all know Hazel? The one who told my ex, Sasha Bernstein, to kill herself in a series of text messages." I stood there, rooted to the spot, as people flooded into the house towards a room I knew all too well. Boos echoed in its walls at the sound of my name.

"It's crazy someone can have that much hate in their body," Sasha's voice said through the microphone. All the hurt I felt in my heart came flooding back painfully. I walked like a zombie towards the room, even though my friends tried to pull me away. "But, mute people have issues, too." I stepped into the room and made eye contact with Grey. It was as if he was waiting for me. He smiled; I wanted to punch his teeth out.

"Well, I just wanted to set the record straight... I always thought Sasha was such a nice person. And she was for some time. But she finally showed her true colors when she cheated on me several times in the past. And what made this whole situation worse..." he said as Sasha watched, mouth slightly agape, "is she lied about, Hazel, constantly."

Surprised murmurs erupted through the crowd, and Sasha looked at him in shock. My rage immediately dissipated. "And you guys like receipts? I've sent them to your phones. And to whoever's number I don't have, spread it like wildfire if you want to. I don't care. I'm just doing my part to clear Hazel's name." He shrugged. "Anyway, remember to have fun and drink plenty of water."

He turned to Sasha and smiled at her. "I'll lead you out." She stormed off the stage as people booed at her. She gave me an angry glance before she left the room. Grey didn't look at me as he followed, his eyes locked on his phone.

Grey sent a text to all of us, and to those who didn't have Grey's number (like my friend group, minus me), it was sent to them by other people they knew. I opened the message and saw screenshots between Sasha and Grey.

Mon, Oct 8, 7:34 pm

Sasha (Ex)I can't stop thinking about what Hazel did today

GreyIt's not a big deal

Sasha (Ex)How can you say it's not a big deal? She threw juice at you!!

GreyClearly I upset her

Sasha (Ex)

Sasha (Ex)Idc, she needs to be put in her place. I found her Wattpad. It's been inactive for years based on her last update, but I found a really corny poem that we could send to everyone in the school. She'll be humiliated it's perfect

GreyWhen I said she'd pay, I was just talking crap in the heat of the moment. I'm not harassing that mute girl

Sasha (Ex)Pft, like you haven't done worse?

GreyOmg.

Sasha (Ex)Stop being so damn self righteous, what is she, your gf?!?!

GreyDude I'm just saying it's not a big deal

Mon, Oct 8, 9:23 pm

Sasha (Ex)You're throwing a party soon right?

GreyYeah why?

Sasha (Ex)You'll see. Whatever I do, you better back me up

GreyWhat are you planning on doing?

Sasha (Ex)It's a surprise

The conversations continued, exposing Sasha for the fake text messages, the wanting to ruin my reputation, everything. Grey had constantly argued with her about it, which didn't make him innocent, but it did at least prove to me that he wasn't actively trying to make my life miserable.

And the most significant thing out of all of this? Grey cleared my name. I beamed as I walked out of the room, my friends following.

"I need to talk to Grey," I signed to them. Ana translated.

"For what?" Andy asked.

"To thank him."

"Thank him?" Blake said and scoffed. "He doesn't deserve any of your appreciation. He's a scum bag still. Just because he cleared your name doesn't make him a good person. He shouldn't have let his psycho girlfriend ruin it in the first place!"

"I need to do this."

"Hazel, I don't think it's a good idea. He can be with Sasha right now, thinking about doing something else that's horrible to you. We should just go home. Please."

"Trust me. I need to do this."

They didn't look convinced, but reluctantly they let me walk off. Right as I did, I got a message from Grey to meet him at the back of the house. Not knowing if Sasha was gone, I cautiously found my way to him. He sat outside over his pool as I slid the sliding glass door open. His legs dangled over the edge of his pool, lit up by the pool's light. "Hey, Hazel." I approached him and pulled my shoes off. I dipped my feet and slowly kicked it back and forth. We sat in silence for a bit, the sound of water sloshing and soft music mixing.

"I'm honestly so sorry for the way I helped Sasha with this drama," he said. "I sincerely hope you can forgive me for being such a... jerk to you." I hesitantly grabbed his hand and gave it a light squeeze. He looked at me, and I smiled warily. He returned it.

"Thank you for fixing things," I signed.

"It was the least I could do. Literally. You deserve so much more than that."

We sat in silence for a little while until I could no longer bear it. Finally, I stood to walk away, but Grey's hand grasped my wrist. My eyes met his, and they were more vulnerable than I'd ever seen them before.

"Hazel. I've had so many dreams with you. I have to know that I am crazy. Because if not, I like you so much more than I'd ever... I... I somehow like you a lot. And I need to know if it's the 'dream you' I like or the real you. Have we dreamed together? Is it possible?"

# CHAPTER 20

"Does it matter?" I signed. My eyes pleaded with his to not push it further. I couldn't keep this act up. And if he ended up knowing, there was so much to be wary about. These questions could be Grey still trying to mess with me for all I knew. I couldn't just give in so easily. I couldn't be that dumb.

"It does, Hazel," he said. He rose and grabbed my hands before I could step away. "If what happened in those dreams are real... if all those things we shared that one dream I had with—about you a few months ago had any truth, then I like you more than I've ever liked anyone before."

He let out a heavy breath and let go of my hands. "It doesn't make sense. I feel like we share dreams, or my mind is obsessed with you. And if it's just my mind, then we have absolutely nothing. And, I don't want that to be true." He looked down at the water. "I've stayed away from you this rest of the school year because I knew I couldn't chase after you when I helped make your life miserable. Sasha hurt you and I let her." His jaw tightened. "So by letting her, I didn't deserve you."

He hesitated for a moment. Loud sounds continued to pour out of the house as a silence spread between us. Grey looked like he was thinking hard in thought. Then his eyes brightened.

"You said..." He didn't finish his sentence. Instead, he lifted my shirt slightly before I could stop him. There was a birthmark on my upper hip that I told him about in the deep talk dream. I was embarrassed by it and explained that's why I hated bikinis. I froze, and so did he. After another shared moment of silence, he let go of my shirt. It fell against my waist. His face was impassive as he looked into my eyes.

"We share dreams." He said it as a matter-of-fact now. "We share dreams."

I couldn't think fast enough. There was no logical excuse I could give. I don't wear revealing clothes, nor do I post on social media often.

Nobody saw my body. Ever.

He took a step closer to me. "Can I kiss you, Hazel?"

My face flushed. I wanted him to kiss me. Very badly. But I couldn't lift my fingers to say so; my hands shook nervously at my side.

He lifted my hand and kissed it softly. It burned where his lips touched. I was so distracted by the feeling they left behind that I didn't realize his lips were on my cheek. They were soft as they gently pressed into my skin. He pulled away, and his fingers gently traced my other cheek.

With a burst of courage, I tilted my head up towards him. His lips met mine moments later. They were soft, and his face was so warm. His feathery breath mingled with mine as our noses pressed together—this was so different from the dreams. Something so real felt so fake and hard to capture. I felt his hands reach for mine, pulling them to his neck. We pressed into each other as our lips danced.

I couldn't get enough of him. I don't think I even ever felt this way about anyone before. Not that I ever dated anyone, but you get the point.

I finally took a moment to pull away. We stayed confined to one another, slightly breathless. I ran my fingers through his hair as he tightened his grip around my waist. We smiled and enjoyed our quiet. I needed him to stay holding me. I couldn't bear the thought of him letting go, even though I knew he'd have to.

I don't know what we were saying to each other wordlessly. Grey's eyes held mine before he placed several feathery kisses on my lips. His hand ran through my hair as the other pulled me even closer to him, which was nearly impossible, but somehow he did it. We stayed there sharing those sweet kisses until a loud voice boomed from in the house, saying, "Grey, bud, where are ya? We need ya to get back to your party!"

"Ugh," he whispered.

I gently pulled away and signed, "You should get back to your party."

"They can wait." He grabbed my hand as he led us to the pool. "This is our time."

***

Grey let me walk back first. I slightly smiled as I walked down the halls, back to the main room. We stayed longer than I thought we would. We sat down at the pool for a bit, and our hands stayed intertwined as he talked. He asked mainly yes and no questions so I could respond to him with a simple nod or shake of the head. I would sign whenever I needed to, but he quickly grabbed my hand again. I eventually began typing on my phone with one hand to talk to him. We could've stayed there all night if I didn't get several texts from the group chat. I eventually left him sitting there as he smiled and watched me walk.

I rounded the hallway into the big room where my friends were standing. They looked at me, relieved.

"How'd it go with him?" Mindy asked. My eyes snapped to her smiling face. She couldn't tell, right? Pulling out my phone, I typed, It went fine. Why do you ask?

"Because it's Grey!" Ana answered as if I was weird for thinking Mindy shouldn't have asked.

He's not all that bad, I wrote.

"Sure," Andy said with an eye roll. "I can attest to his niceness."

"Are we done with this party yet or..." Blake said, yawning.

"I think we are," Mindy concurred.

We walked through the mob of kids until we got to the front door. Before I could leave, a hand grabbed me and pulled me back into the house. I looked up and met a familiar pair of brown eyes.

"Wait," he begged. I couldn't give a response before his lips were on mine again. He held me so close to him—I could feel the strength of his arms, the heat from his body, the softness of his lips. I tried to memorize it all.

He pulled away, and I looked up at him with probably the most apparent blush coloring my face. My eyes hesitantly glanced around—most of the kids were too drunk even to notice the exchange between us, and my friends hadn't realized I was missing yet. He smiled as I reached up for his neck and stole another kiss.

"I'll text you," he said, and I nodded. We let go of each other, and I ran out of the house. My friends still didn't notice I was missing yet. I glanced back at the mansion once more before hopping into Mindy's car.

***

GreyTomorrow, let's go out on a date. Okay?

MeSounds good. When and where?

GreyI'm going to take you to that roller skating rink in town. You okay with that?

MeThat sounds great.

GreyI'll pick you up around three?

MeYeah

We spent the rest of the night talking until I eventually fell asleep with my phone still in my hand.

***

Discomfort spread throughout my chest, but I'd be lying if I said I didn't love it once I realized I was in Grey's dream. He sat down on the shoreline of a white sand beach. The water lazed up the shoreline toward him, then returned to the frothy ocean. He leaned on his elbows, enjoying the ebb and flow of the water.

I sneakily made my way over to him before tackling him into the sand. He let out a startled "whoa!" before eating a face full.

"Oh my gosh, I'm so sorry!" I laughed while falling off of him. He pushed himself up and playfully glared at me.

"I should make you regret that," he said.

"Oh, please do," I said sarcastically. His fingers grazed my arm before he gently caressed my face.

"How come you didn't tell me we shared dreams?" he asked. "I felt like I was insane the way I had such unexplainable strong feelings for you."

"Because it's not normal," I said.

"It's amazing!" He tickled me, and I giggled. "You have such a cool gift."

"Thank you," I said shyly. "But if people knew I was in their dreams with some of the things I've seen... I can only imagine the consequences."

Grey's eyes widened in understanding. "So you like to keep it a secret so nobody can use it against you."

"Yeah, people either use it against me or think I'm crazy."

"I understand." His fingers intertwined with mine. "Your secret is safe with us."

I smiled. "Us?"

"I'm tired of running from my feelings. I want us to be official."

"So why don't you 'officially' ask me out tomorrow?"

"I surely will," he said. "Or you could ask me out, you know." He nudged me teasingly.

"I could," I said. "But, I don't wanna. You owe me."

"I owe you more than I'll ever be able to repay," he said, slightly seriously.

"It's okay," I said, and my fingers caressed his hand. He pressed a firm kiss on my lips and then laid down next to me on the sand.

"Wanna go swim with sharks that won't try to eat us?" he asked.

"Heck yes!"

# CHAPTER 21

I woke up smiling.

I loved being in my lucid dreams because those were fun. But being in a lucid dream with someone aware they were dreaming and knew they shared it with me was amazing.

I was awake for maybe five minutes recounting the dream when I got a text from Grey.

GreyDid we have the same dream?

MeWhat happened in yours?

GreyWe held onto a great white shark's fin until we floated over an underwater city

Meand then we were greeted by a creature that was a cross between a starfish and a turtle?

GreyYes! So much happened it was so much fun. That is honestly so cool

It warmed my heart to read those words, as meaningless as they may have been to him. No one else has ever shared something like this with me. And the fact that I was able to make it enjoyable for him made it even better. Before I could respond, My mom knocked on my door. Startled, I closed my phone and sat up. She smiled at me, surprised. "Oh, you're awake! I'm going grocery shopping. Did you want anything?"

"No," I signed. "Thank you."

"All right, I'll be back soon," she said and shut the door.

I unlocked my phone again to see Grey's new messages.

GreyReady for our date later today?

MeYes!!! I'm so excited

GreyI'm glad. I'm nervous lolol

MeGrey and nervous don't seem like they fit together

GreyThen you and I have a lot to learn about each other

MeI'm very happy about that fact

GreySame. I have to make up for a lot of crap that I've done btw

MeI'll hold you to that

GreyYou should send me your address okay? I gotta go help my dad with some stuff for my sister (she's staying over in California with my aunt and uncle over the some instead of coming back from her boarding school), but I'll call you when I'm getting ready to get you. Cool??

MeSounds good!

Grey's sister! I never got to see her because she was still an unborn child with whom I learned of her existence. I hoped to meet her soon, though.

My nerves ran wild as I waited for Grey. I scrutinized my outfit several times, gave Timmy excited belly rubs which would rile him up—he'd run off and grab his favorite toy, growling at me to take it from him as he ran away with it—and laid in bed with excited butterflies filling my tummy.

When I got the ping of his ringtone letting me know he was downstairs, I practically ran out of the house. If it weren't for my mom being back from grocery shopping, we'd be in the car, happily flooring it down the street.

"Baby... you didn't tell me your plans...?" my mom said from the kitchen as I slipped my shoes on. My happiness came screeching to a halt. My mom—wasn't strictly okay with me having a boyfriend. If I recall correctly, she said once I was in college and out of her hair, that's when I could think about dating.

Well, at that moment, Grey and I were still friends—friends who were going on a date together.

Yikes.

"Going to the roller skating rink," I signed.

"With?" she asked. My palms grew sweaty. I didn't want to lie to my mom—that would only make things worse. But at the same time, she would make my life a lot more difficult if I flat out told her the truth.

"A friend," I signed.

"A friend?" Her eyebrows raised. "Does this friend have a name?"

"Grey."

Her mouth flattened, and she sat down at the kitchen table, signaling for me to do the same. "Is this a date, Hazel?" I sat down hesitantly and signed, "Sort of."

"You know that I don't want you dating, Hazel."

I stared down at the table, my heart crashing and burning from her words. She sighed after a minute and said, "I'll let you give this a go. If this relationship distracts you from school, you're both done. Got it?" I nodded solemnly and rose from the table. She followed me to the front door, and I gave her a questioning look. "You think I'm not going to meet this kid?" I smiled warily.

Grey stood outside his car, looking as handsome as ever. It was weird to think he was here for me, wearing a happy smile, staring longingly at me.

"Hey," he said, smiling. His eyes met my mother's, and he smiled at her. "Hi, I'm Grey." He pushed himself off the car and outstretched his arm to give her a handshake. She returned the smile.

"Nice to meet you. I'm Hazel's mom." She gave me a slight smirk. "I'll meet you more officially later. Don't stay out too late. I know it's summer, but I'm a worried mother who will come to beat you both. Have fun." With that, she walked back into the house.

"I'm important enough to meet the family, hm?" he said with a smile full of glee.

"My mom is overprotective. Of course, she'd want to meet you right away." He wouldn't stop smiling at me. It made my heart swell over average capacity.

"What?"

"I'm just... happy to be here with you." He wrapped his arms around me in a tight embrace before asking, "Ready?" I nodded.

He opened the door for me and then hopped on the other side. "What would you like to listen to?"

"Up to you," I signed.

"My choice is yours," he said.

I pulled out my phone to pick out a song. Then he connected my phone to his car, and we cruised down the street with the music blasting.

"I like this song," he said, nodding his head to the rhythm. I smiled and started bobbing my head along with him. He glanced over at me and smiled. "You're super cute."

My cheeks heated up, and I poked him in the arm.

For the rest of the ride, he cracked jokes that got me to laugh. He also talked more about the dreams we've shared, including ones he didn't realize were because of me.

"That dream where you pretty much helped me get over my fear of dogs was the first time I started to feel something for you." I smiled as I stared out at the road. "I didn't know what I felt for you then, though. Sasha was still able to get to me." I nodded, her name making me queasy. "Seeing her sad was enough to make me want to go back. And I knew the consequences. I still made my choice, but I wouldn't have done it if I had known it'd burn you. She wouldn't have stepped in the way she did if I hadn't gotten back with her. She would've said I deserved it."

I stayed silent, knowing I'd have to talk to him more about this in our dreams later that night.

We pulled into the parking lot and hopped out of the car. Grey walked close to me and gingerly took my hand. I responded by squeezing. Hand in hand, we made our way inside the building. Without even discussing it with me, he automatically paid for my ticket.

"Grey!" I signed. "You don't have to do that."

"Sh," he said and kissed my cheek. I couldn't help the smile that appeared on my face after.

Once we had our roller skates on, Grey could tell I didn't know what I was doing.

"You okay?" he said, and I nodded sheepishly. "Do you need me to hold your hand, darling?" His voice was teasing as he lifted his eyebrows up and down repeatedly.

"Maybe," I signed and slightly lost balance. He caught and steadied me. He rolled us over to the rink. Holding hands, we began to skate together. The heat from his hand flooded through my body and made me feel so warm. I wonder if he was honestly even nervous about being with me like this. His hands weren't sweaty like mine. He seemed so calm and collected while I, on the other hand, was not. I shouldn't have been nervous after the times we've spent together at the pool and the number of times we shared dreams, but the mind was a funny place.

"Look, you're already a pro," he said.

I gave him a look that he was able to read immediately.

"I'm not just saying that," he said. "You're doing a good job. Remember to keep your legs slightly bent, and your upper body leaned forward rather than back." I did as he said, and it made my balance better.

He let go of me and let me go off on my own slightly. I reached out for him desperately.

"I'm right here," he said soothingly. "Keep going. You're on a roll. Ba-dum tsk." I laughed and lost balance. He caught me with ease and rolled us back to the wall. He pressed me between the wall and him. He kissed my nose before letting me go sliding beside me.

"You almost killed me," I signed playfully.

"At least you would've died laughing," he said with a slight smirk.

"Your joke was dumb," I signed, hardly containing my laughter.

"I am hysterical," he teased.

"Oh yes, and very humble."

"What can I say," he said with a sly smile. His hand reached for my cheek, but something made me look around. My blood ran cold. Sasha was on the other side of the rink, her eyes fiery but calm. She smiled as if she knew something I didn't.

My heart slammed in my chest. Grey set me up, didn't he? His whole plan was to get me here and have to face Sasha. Why else would she be here alone? I felt sick to my stomach.

I looked up at him, hurt. He looked back at me, confused, and then to where I had been looking. His eyes met Sasha's.

# CHAPTER 22

Grey looked back at me, shaking his head. "Hazel, I swear, I have no idea why she's here." He glared at her, and she frowned in response.

"Let's go," he said, grabbing my hand. "We're gonna leave before this turns into something." We skated against traffic and exited onto the main floor, following his lead.

"Grey," Sasha called in a sing-song voice. His back tensed up, and he pushed me in front of him before facing her.

"What do you want, Sasha?" he asked as politely as possible.

"I can't say hi to my friends?" she asked and then smiled a bright, fake smile. "How are you doing, Mutey?"

"Why are you still calling her that?" he asked.

"It's my cute nickname for her," she said casually.

"Well, she doesn't like it. So knock it off."

She rolled her eyes. "So I see you two are buddy-buddy now." My heart slammed in my chest as she glared at me. Why the heck did she hate me so much? I did nothing to her. I threw that juice on Grey. Why did she take such high offense? I get defending your boyfriend but to this extent? It was ridiculous! And she wasn't even with him anymore. Was I? Technically? I didn't even know at this point. I was still scared that this would turn on me somehow, and Grey would laugh in my face and tease me alongside Sasha.

But he didn't. He remained between Sasha and me as she continued to speak. "Are you guys dating?"

"That's none of your business."

She scoffed. "Hazel, just to warn you, honey, he doesn't like you. He just wants to sleep with you and then try to come crawling back. As he always does."

"That's rich coming from the cheater," he said cooly as she gave him a perfect smile. "I never did anything when you slept around with Evan." He leaned in closer, never losing his balance. "I should've, though. You most definitely deserved it."

Her eyes flashed with hurt. "This is the type of guy you want to get with? Someone who thinks that cheating is an acceptable way to get back at someone else?"

I glared at her and remained silent. That was all I could do, anyway.

"If looks could kill," she teased. "You're so cute when you're angry but silent. Don't you have any words for me?"

"What's your obsession, Sasha?" Grey asked calmly. "You're like a dog with a bone. You don't know how to let go. Get over yourself and leave us alone."

"Get over myself?" she asked, her voice shrill.

We didn't even see Evan approach, but he stood behind Sasha and put his arm on her shoulder. "Sasha, babe, come on." He looked between Grey and me apologetically. "Let's have fun."

"I would've had fun if it weren't for Grey and his Mute bi—"

"Sasha," Evan snapped. "Let's go."

He was quickly able to yank her away from us as she protested angrily. "Who's side are you on, anyway?! She deserves it!"

The tension left Grey's body, and he slowly turned to face me. I hadn't even realized there were tears in my eyes flowing down my cheeks until he wiped them away. "Don't let her get to you," he said and kissed my forehead. "She's not worth it. She just wants to get under your skin."

"I don't get why she hates me so much," I signed. "I didn't do anything to her."

"Even when I was with her, I didn't understand it. I just backed her up because I was trying to keep her. I was an idiot for that."

I nodded. Grey was. He was so stupid for ever supporting someone so hateful. It hurt to know that he helped make my life miserable.

But he also cleared my name, I reminded myself. So if he cleared my name, he knew what he was doing was wrong and didn't fully support her. He had a mind of his own and a conscience. And he finally listened to it.

I grabbed his hands and wrapped them around my waist, pressing my head into his chest. He leaned his head on top of mine. We breathed together there, and I felt the anxiety washing out of me. He was with me now. Not her. He was finally on my side. He was finally holding me and finally kissing me. It was no longer a dream but an actual reality.

"I guess this is a bad time to ask if you'd wanna date," he said. "But you already know I like you, Hazel. And I told you in our dream that I would ask you. Can we make this official?"

I pulled away, and I nodded with a bright smile on my face. Grey grinned, and his lips met mine softly. He repeatedly rubbed his nose against mine, causing me to smile and squeeze him. He chuckled.

"You're so cute," he said. I kissed him in response.

We eventually let go of each other and traded in our skates for our shoes again. We hardly got to have fun before Sasha ruined it, but I got something good out of it. Grey made me feel a little bit better about where we stood. Especially now that we were official. I looked at him appreciatively in the car as he drove us to get some food. He looked back at me with a slight smile on his face.

"I know I'm handsome, but dang, I guess I'm more handsome than I knew," he said, "since you're staring at me for so long. Soak it all in, Hazel."

I laughed at his playful arrogance and poked his leg.

After eating food and kissing me on a park bench, Grey finally decided to get me home. He pulled up to my house, and I gave him a quick peck before getting out of the car.

"See you later," he said with a wink. I smiled knowingly.

Grey pulled away from my house once I unlocked my door and made it safely inside. My mom was waiting for me in the living room with Timmy curled up beside her on the couch.

"How'd your date go?" she asked.

"It was nice," I signed. "We skated for a bit, got food, and then sat in the park."

"Was this park public?" she asked.

"Yes, mom," I signed. She smiled. "Good girl. Now, tell me about this boy! I wanna know all about him."

I couldn't tell her much. Not without it blending into what I know about from our dreams. The truth was, I knew Grey more from being in our dreams together than in the real world. That was why we were getting to know each other now.

"He's sweet," I signed. "He was rough around the edges when I first met him, but he softened up after a while. I developed feelings for him, and he felt the same way. I don't know, mom." I was getting shy. "I just know I like him a lot."

"I can see that," she said. "My only rule is no sleeping with him. Got it? I feel like I'm letting you get away with murder at this point."

"I don't want to sleep with him, mom," I signed, horrified.

"Maybe not yet," she said and left it at that. "Wanna watch a movie?"

"Sure," I signed and plopped down. "What movie?"

"Jurassic Park?"

I nodded. It was one of my favorites. I sat close beside her and leaned my head into her shoulder. She kissed the top of my head as she pressed play.

***

Grey met me inside a diner as the rain poured outside. I had a cup of coffee in my hand, and I took a sip as he sat down across from me in the booth. The seats were red and cushioned, and the table was wooden and small. Our knees pressed together as he smiled at me.

"Do you feel uncomfortable?" Grey asked. "It feels a little weird in here."

"It's because you're in my dream," I commented. He smiled at the sound of my voice. "So you feel like you don't belong."

"So you can take people into your dreams?" he asked. "That's a part of your cool abilities?"

"I don't know," I said and shrugged slightly. "This is new to me. I fell asleep expecting to be in your dream."

"Well, we're here together, and that's fine with me," he said. He took my hand and pressed a kiss into it.

"Can I admit something to you?" I asked quietly. He squeezed my hand as he nodded. "Of course. What's up?"

"When Sasha was there at the skating rink... I thought you set me up. I thought you were both about to humiliate me somehow."

I watched as his face grew solemn. His thumb caressed my hand as he said, "I could tell seeing her bothered you. I didn't realize you thought it was because I was setting you up. I guess you don't trust me."

"Not as much as I want to. Not yet." I gave him a half-hearted smile. "It's a little hard to after all that happened between us."

"Yeah," he said and sighed, releasing my hand. He clasped his fingers together and leaned his face into them. "What can I do to change that?"

"I think that'll just be something that'll take time." A waiter walked over to us with a pitcher of water. Grey accepted and allowed him to fill up his glass. I shook my head politely. The waiter walked away as Grey took a sip. We sat there in silence as the rain continued to pour outside.

"I guess that's true," he said finally. "Let me know if there's any way to help you not worry as much."

"I mean... there is. Do you still like her?"

He laughed. "Hell no. Sasha and I had a weird relationship." He looked out the window and tapped his fingers against the table. "Sometimes it was perfect, and other times I didn't know why the h-e-double hockey sticks I was still with her. Part of me thinks it's because she reminded me of my mom. Her good aspects, at least. And, disclaimer, I'm not perfect either, I know that. I have a lot of crap to sort out, and we just couldn't handle each other's crazy." He looked at me dead in the eyes. "But I assure you Sasha is nothing to me now. She's in the past and will stay in the past. And I know that's something I'll have to prove to you by my actions. But, I still want to put that out there."

"Thank you," I said shyly. I smiled, and Grey smiled with me. I intertwined my fingers with his. "Y'know... you weren't the only one affected by these dreams."

"Oh?" He looked interested and dropped his guard. He nudged me under the table with his leg and wore a tiny smile. "So I wasn't alone in all my feelings for you?"

"I probably wouldn't be with you if that were the case," I teased. "If it weren't for the dreams, I wouldn't have known that you were a nice person underneath all that tough guy act."

"It's not an act," he said with a small laugh. "I don't like people like that, and teachers and our classmates don't get to me. But, for some reason, that got me popular. I don't even understand the hype." He shrugged. "I guess it's a good thing, though. I don't have to deal with people trying to mess with me. They already know the deal."

"You seemed nice to me initially, though," I pointed out.

"Yeah, my dad taught me to be nice to females." He looked embarrassed to say that. "I'm sure if he heard what happened between me, you, and Sasha, he'd be disappointed in me."

"We all do things that disappoint our parents," I said. "Don't beat yourself up about it."

"I'm not. I'm holding myself to a higher standard. And you better hold me to that standard as well."

"I will," I promised. "And you deserve to be treated right by me as well."

"And that's on non-toxic relationships," he said, causing us to laugh.

We heard a roar in the distance, causing both to freeze. Stomps followed after it.

"What is going on?"

"I may or may not have watched Jurassic Park before this," I said sheepishly.

"Is this going to be terrifying?" he asked.

"Yes," I said honestly.

"I will never forgive you," he said as a T-Rex's eye and face appeared in the window next to us.

I laughed. "I'm just playing. I wouldn't do that to you." Then I shouted, "Wake up!"

# CHAPTER 23

The next day, Grey took me on another date to the movies.

I frantically searched for another cute outfit. I shook out my hair and fluffed it slightly to look good. My waves had seen better days, but Day 3 hair was never good for me. I checked my outfit out several times in the mirror and changed it countless more times. Finally, I was slightly satisfied with the look I pulled off.

I ran downstairs and sat in the kitchen. Scrolling through her phone, my mom looked up at me and smirked. "You look cute. What time is the movie?"

"At five," I signed. "We're gonna get there a bit early."

"Gotta catch all those previews," she said.

"Exactly."

I sat down across from her. "What have you been up to this week?"

"Work," she said. "The same ole, same ole. This mundane life is what you get to look forward to." She shook her head. "It ain't all fun and games, kid. Enjoy your youth now."

"I do," I signed. "I'm not in a rush to grow up." Even though I felt like I aged years after witnessing my dad's death, I wouldn't bring that up for no reason.

"Mhm." She looked at her phone again, scrolling mindlessly before looking up at me again. "So, are you and this boy serious about each other?"

"I think so," I signed.

"Think so?" she questioned.

"I haven't let my guard down."

"Why would your guard be up?"

Oh, crap. I didn't tell my mom about any of the drama going on, nor did I plan to. If she knew Grey was part of my life, making me miserable, she would disapprove of him. He'd automatically be someone she'd think didn't deserve to be anywhere near me. It was a little more complicated than how she'd see it.

"I'm not going to just go into a relationship and automatically trust him," I signed. "Shouldn't we be earning each other's trust?"

She held my eyes for too long of a moment. "You better not be letting some little heartbreaker gain access to your heart."

I laughed. "No, Grey's not like that."

"Good," she said firmly. "I will not tolerate boys who think it's okay to mess with girls' feelings. Boys are stupid, anyway."

My phone buzzed, and I pulled it out of my pocket to see a message from Grey letting me know he was outside. I stood and kissed my mom goodbye.

"Love you, baby, see you later. Be careful!"

"I will," I signed. "I love you, too."

I kissed Timmy's head before running out of the house. Grey sat inside the car. I pulled on the handle and sat in the passenger seat.

"Hey," he said happily.

"Hi," I signed.

"Buckle up."

We drove in comfortable silence, listening to music on the radio. I would glance at Grey and catch him doing the same thing several times. We smiled at each other before looking back at the road.

He didn't make me feel bad that I couldn't talk. The quiet didn't seem to bother him at all. That was something that crossed my mind several times. Would he care that I couldn't say anything? But, so0 far, it didn't seem like he minded it.

We pulled into the movie theater parking lot. It was crowded. Grey lowered the music as he scoured around for any empty spaces.

"Look at this moron," Grey said with a sigh. "Taking up two parking spaces. This is why people should take an IQ test before driving." I laughed. He finally pulled into a spot, and we exited the car. He met me behind the car with his hand already extended. I took it happily, and we walked toward the movie theater.

Once inside, we got on the popcorn line and stood close to each other.

"What do you want?" he asked.

"Popcorn," I signed. Grey looked at me amused and said, "I figured, but what size do you want?"

"A small," I signed shyly.

"Okay, so a large," he said.

"No!" I signed. "I honestly can't eat that much popcorn."

"More for me," he said with a wink. "We'll share it."

I rolled my eyes with a smile and nodded. We got to the counter, and Grey began to order for us. That's when I heard a surprised gasp behind me. I spun around to meet Ana and Blake's wide eyes staring back at me.

Grey turned around, too, with the popcorn, various snacks, and drinks in hand. The poor guy had no clue that my friends didn't like him, although I'm sure he could guess from the daggers they were sending his way. I looked at him, and he gave me a tense smile. "I'll go save us our seats."

He walked off, leaving me to face my friends' wrath alone.

"Hazel, why are you here with Grey?" Ana exclaimed. "Have you forgotten what a complete a-hole he is?"

"I wouldn't be dating him if he was," I signed.

"If he was?" Blake said. "Hazel, he is. You can't have forgotten that quickly how badly he treated you."

"I promise you don't know him how I do," I signed, and Ana scoffed.

"Don't even say that," Ana said. "Please, girl, take those rose-colored glasses off your eyes and see him as he is."

"They were off when I started liking him," I signed. "You need to give him a chance before writing him off!"

"A chance to hurt our sister?" she asked. "You think we'd stand there and let that happen even though you tried to hide it from us?"

"So you want to talk about me being your sister and hiding stuff, huh?" I signed. "Why are you two here on a date when you didn't even tell me you liked each other?"

Their mouths closed immediately. Their eyes met guiltily before looking back toward me. "Hazel, of course, we wanted to tell you," Blake said. "It's just that if it doesn't work out between us, we didn't want you to feel awkward and torn between us. We rather you'd be clueless instead."

So, the reasoning finally came out. I wondered why And and Blake wouldn't share what they had together with me all this time, especially when I was so happy for them. But it was a pretty good train of thought—if I didn't know and they ended up having a falling out, they wanted it to be that it wouldn't fall on me. After all, we were a trio; we were beyond family.

But I would still know. I'd feel their distance from each other. I'd see the hurt in their eyes every time their eyes met. And it wouldn't take me long to figure out what happened.

"I get that," I signed. "But just like you didn't like me hiding this from you, you can understand why I felt the same way." They nodded.

"Are you going to theater two as well?" I signed.

"Yeah," Ana said.

"How about we watch the movie and get dinner afterward?" I suggested.

"That's a good idea," Blake said. "We get to interrogate the heck out of him." I laughed, and the tension left my body.

We walked to the theater together as they teased me.

"Have you guys kissed already?" Ana asked. My face flushed, and my ears turned red.

"Oh, I'll kill him," Blake said. "I'll wring his neck!"

"Be quiet," I signed.

"You're so stupid, Blake," Ana said with a laugh.

"The nerve of this guy thinking he can do more than hold her hand!"

"He's right, Hazel," Ana said. "No kissing is allowed."

"Then you two aren't allowed to kiss," I signed.

"Heck no," Blake said and kissed Ana quickly.

"If you guys weren't holding drinks, I'd attack you."

We walked into the theater. Grey was seated a few rows up, three other seats next to him saved for us. I couldn't contain my smile.

I sat down next to him, and Ana and Blake sat beside me.

"Thanks for saving our seats," I signed.

"I figured they'd wanna join us," Grey said. "But you're welcome." He kissed me lightly, and I felt a slight kick on my leg. I held in my laughter until we pulled away.

"No kissing," Ana whispered, and I giggled while kicking her back.

We talked for a bit more, mainly Ana and me, until the movie started. As we settled in, Grey's arm went around my shoulders.

"They don't like me, I'm guessing?"

I pulled out my phone, facing him, and wrote, They're still unsure about you. Give them time. You mind if we have dinner with them after the movie?

"Sounds good to me," Grey said and kissed my temple.

***

Grey sat next to me, and Ana and Blake sat across from us after getting our food from Wendys. Paper crumpled as we unraveled our sandwiches. We were all still very uncomfortable now that we were all together. I squeezed Grey's leg and tried to give him my happiest smile. I could tell he was nervous as we drove over here. He repeatedly drummed his fingers against his wheel and leaned his head into his arm when we stopped at any red lights. I couldn't even grab either of his hands to reassure him.

For someone who claimed he didn't care about people, he cared about how Blake and Ana were going to view him.

"So," Blake said, "how long have you two been dating?"

"Two days," Grey responded. "How about you guys?"

Ana looked at me warily. "Since the beginning of the junior year."

Grey whistled, impressed. I was, too, although I also felt a slight sting. But I couldn't be angry at them anymore. They explained why they hid it from me, and I understood it. "So, you guys made it past the six-month mark. That's nice."

"So, do you like Hazel?" Ana asked boldly. "Because after all that you and Sasha have done, I don't think any of us trust you."

"Yeah, I know," Grey said and let out a heavy exhale. "I like Hazel a lot."

"Why?" Blake said.

"Guys, c'mon," I signed.

"It is a little personal," Grey said. "But honestly, what's not to love about her? Aren't you guys protective over her for those reasons?"

I bit into my burger, feeling warm and uncomfortable. A lot of attention was put on me even though they weren't looking at me. So, I decided to change the subject.

"Grey likes to cook; isn't that cool?" Such pathetic wording, but it managed to get the ball rolling.

"What do you like to cook?" Ana asked.

"Anything I can, actually," Blake said, his eyes lit up. "There was this cool Kimchi ramen recipe that I saw online that I wanted to make. And also these kimchi spam quesadilla's I've been meaning to make for my dad and me. They look delicious. But yeah, I like to cook because of my mom."

"Oh, so did she teach you how to cook or...?" Blake asked.

"Not really," Grey said. "I like to cook because it reminds me of how she'd cook all the time."

"Oh, so you do the cooking in the house now?" Ana asked.

I saw where this was going, but Grey didn't mind. "Yeah. My mom passed away some time ago. I cook even though I don't need to."

"Oh," Ana said, sitting back slightly. "I'm sorry to hear that."

"It's all right," Grey said with a shrug. "We miss her every day, but it no longer hurts as bad. I guess you get used to it after a while, and we've overcome it." Blake and Ana didn't look like they knew what to say, so Grey helped them out. "So, since we're doing this awkward small talk," he said, placing his palms on the table and leaning his head to the side, "what are your guys' hobbies?" That got them to laugh.

"Breathing," Ana joked.

"Ah, yes, my favorite," Grey said and then looked at Blake. "And would yours be living?"

"That's my second favorite hobby," he said. "My first would be cleaning. I'm a neat freak which both Ana and Hazel hate." I nodded vigorously.

The ice slowly melted at the table as they continued to talk. My heart swelled in happiness. They were getting along nicely.

We lost track of time sitting there. When I finally checked my phone and saw a "Where are you?" text from my mom, I had to sign, "Oh, I've got to get home!"

I quickly texted her, On my way, sorryyyyy. Blake, Ana, and Grey were talking, and we lost track of time.

She responded, It's okay, just come home soon.

I hugged Ana and Blake goodbye in the parking lot. Ana whispered in my ear, "He's passing so far."

"I'm glad," I signed when we pulled away. "Get home safe, okay?" They nodded, and Blake said, "You too, text us when you're home."

I grabbed Grey's hand, and we walked to his car and hopped in. As he drove me home, I could see his mood improved drastically. He was calmer, and he had a content smile on his face. I kissed his hand, and he kissed mine back.

He pulled up to my house and said, "That went so much better than expected." He put the car in park and faced me. "Do you think they like me better now?"

"Definitely," I signed. "They'll still probably be cautious around you, but you sure got them to feel more at ease."

"I'm glad for that," he said. "I had fun even though our date didn't go as planned."

"Me too," I said.

He leaned over, cupped my face in his hands, and kissed me for a long moment. "See you later tonight."

"See you."

# CHAPTER 24

I felt discomfort as I opened my eyes. A plain-looking three-story house sat in front of me. The shingles on the roof fell to the ground as I approached the building with plain white walls. Something compelled me to press the doorbell.

Part of me was afraid that I somehow got caught in dad's murderer's dream again. If he opened the door, would he even recognize me? Did we ever look at each other? I didn't know. I know I never saw his face. Who knows if he saw mine before he grabbed me?

Much to my surprise and also my dismay, Sasha opened the door. Her eyes went wide, and I'm sure mine did, too.

"What are you doing here?" she asked.

"I don't know," I replied. Sasha didn't seem surprised that I was speaking.

She took a step out of her house and closed the door behind her. I could already tell she was hostile. I began to walk back down the steps, and she followed me closely.

"Did you come here to rub it in, Hazel?" she hissed.

"Rub, what in?" I asked, seriously confused.

"The fact that you were able to get Grey wrapped around your stupid, little fingers?"

I narrowed my eyes and shook my head. "Why would I do that?"

"Why wouldn't you?" She stopped walking, and so did I, ensuring we maintained a reasonable distance from each other. The wind howled around us and blew her hair into her face. She angrily brushed it away as she glared at me. "You want to gloat 'cause you took him away from me."

"I took him away from you?" I practically shouted. "You cheated on him several times before I even knew him. I didn't do anything that wasn't already going to happen eventually. Even if I weren't involved, he would've left you."

"There was a reason I cheated so much," Sasha said defensively. "You wouldn't understand."

"Try me."

She regarded me cooly and then smoothed her hair back. "Grey never loved me or even liked me the way he likes you. He hardly gave me the attention I wanted. And nothing I did seemed to work until I cheated. Then he was all over me." She stomped her foot. "But he got distant again after a while until I cheated with Evan. And Evan was the only one who continuously gave me attention after we slept with each other. And that drove Grey crazy." I felt sick to my stomach.

"You're both trashy people."

Her eyes held daggers, and she crossed her arms. "I told you you wouldn't understand." The wind around us began to blow harder as her face hardened even more.

"I understand it. But it doesn't justify your actions at all. You should've broken up with Grey if you felt you could get treated better somewhere else."

"I loved him," Sasha protested through grit teeth. "I didn't want to leave him."

"You loved him?" I laughed humorlessly. Enraged, I picked up dirt from the ground and molded it in my hand. I needed her to be quiet before I lost my cool. "That's a joke, right?"

"I did love him, you b—" In anger, I smeared the dirt across her mouth. It acted like glue. Her eyes narrowed as she repeatedly tried to wipe it off. She looked semi-panicked when she finally looked back at me, probably realizing it wouldn't be coming off.

"You didn't love him," I stated. "If you did, you wouldn't have cheated. And instead of acting grown, you behaved like the spoiled brat you are and threw a fit until you could get your way. And look where that got you!"

There was a fire in her eyes. She tried to shove me, and I quickly stepped away from her. She fell to the ground and screamed out angrily. She stumbled up and immediately swung. This altercation was ridiculous, and I wouldn't entertain it anymore. I slowed down her movements as I stepped away from her and floated into the air. She watched me as I went, not noticing the heavy footsteps approaching from behind.

When she finally felt them, she stiffened. She slowly turned around to see a T-Rex towering over her. Her brain pushed her further away from it before she turned and started running. I felt the tug of wakefulness retaking hold of my brain. The T-Rex chased after her slowly from behind. She looked around for ways to hide as the T-Rex got closer. Her subconscious fought against mine as I prevented any convenient hiding spots from popping up.

It roared behind her; she fell to the ground in fear, letting out a muffled cry for help.

I decided she had enough and dropped down beside her. The T-Rex nudged its head into the palm of my hand, and I whispered, "Thank you." It walked off while I looked down at her.

Tears were in her eyes. I wiped the mud off her mouth and tried not to feel guilty. Weakly, she stood up and said, "Where did that thing even come from?"

"I don't know," I said with a shrug. She watched it walk-off from over my shoulder before looking back at me. "Why do you hate me so much, Sasha?"

She sucked in her cheeks before rolling her eyes and lifting her chin. "I'm jealous of you. Even when I read that crappy poem about your father's death, I could already tell he loved you more than my father ever did." I sighed. I should've seen this coming even though I didn't want it to be the case—so that I wouldn't feel bad for hating her. But, of course, there was some reason she was acting the way she was. It wasn't because she was a genuinely bad person. It was because something was going on in her life.

"Why do you think your father doesn't love you?" I asked.

"I know he doesn't," she said. "He's hardly around. And when he is home, he never gives my mom or me any attention. He doesn't love us."

I didn't know what to say. "I'm sorry, Sasha."

"Spare me," she said as she wiped tears from her eyes. "The only way my mom got attention from my dad was when she started hanging out with her male co-workers. When my dad found out, he ended that real quick. My mom couldn't go out in public for a while. And then he threatened to beat me the same way he beat her if I ever did anything as stupid as that to a man. So I did it anyway because he wouldn't do anything, and I was right. He didn't do crap because he never cared."

She crossed her arms again and stared down at the ground. "Same thing for Grey. He either didn't love me or fell out of love with me. Whenever I wanted to do something with him, he was busy with the boys. I wanted to go places with him, but he was tired of them. I wanted to have Us Time, and he said he couldn't come over. He used to want to spend time with me always. Then it was just him trying to do stuff he wanted to do instead of what I wanted to do.

"The last time we got back together, I could tell it wasn't out of love. And when he tried being kind to you even though you deserved what was coming, I felt broken. He shouldn't have cared about you in any way. He shouldn't have cared about any other girl except me."

"Sasha, Grey cared about you very much," I said. "You should've talked to him. And if he didn't give you the attention you needed, you should've dumped him and found someone who would. Like Evan. You went about it the wrong way, and now people only see that bad side to what you did."

"But you can see the other side?" she asked numbly.

"Yeah," I said. "It doesn't mean I think it was okay."

"I know," she sniffled. "Whatever." She walked off to her house, which was much closer than when she ran off, and I slowly followed behind her. She walked up the steps before stopping to look up at me.

"I don't regret trying to hurt you," she said. "I never will."

Of course not. I imagined a hole on the last step. "Wake up for me, will ya?" She rose her eyebrow before turning around and stepping into the hole with a scream.

***

I found myself back into a familiar discomfort. It was funny how Grey's subconscious started to make me feel at home, although I didn't belong in it.

"Grey," I called. He was sitting down on a park bench. There was a sunlit open meadow in front of it with various animals playing around in the greenery.

He waved at me and said, "Hey! I've been waiting for you."

"Sorry I took so long," I said, throwing myself down next to him. "I just came from Sasha's dream."

"What?" His interest automatically peaked. "How did that go?"

"She tried to attack me," I said bluntly. "But we also had a very interesting heart to heart. It was friggin weird. People are so erratic when they don't realize they're dreaming."

He held one of my hands in his and said, "Tell me about it."

"I'm not gonna tell you too much because it was very personal for her," I said. "But she tried to justify cheating on you because it seemed to be the only way she'd get any attention from you."

"What?" His face contorted in anger. "I gave her so much attention. I only distanced myself because nothing I did seemed good enough for her, and then she started cheating on me, so I didn't want to be around her."

"I told her she should've just spoken to you," I said. "I felt bad for her because of the stuff she told me about."

"What, with her father?" he questioned. I nodded as I said, "Yeah."

"Her father was a jerk, no offense to him," he commented. "But I did try to be nothing like him. It felt like we always did something she wanted to do and nothing we wanted to do. I wanted to go to this mall one time where we could both find the stuff we liked, but she just wanted to go to a particular store that only had girls' clothes. I told her, 'Okay, you have a car, and I have a car, let's get our stuff and meet later on in the day. She got angry at me for that. So we stayed in this loop where I didn't feel good enough for her, and she didn't feel like I loved her."

"Yeah," I said. "I'm sorry."

"I never meant to make her feel that way, though," he said and looked down on the floor. "I feel bad that I did."

"It sounds like you guys should've talked everything out."

"Probably," he said.

"Do you want to try again with her?" I asked. He looked at me in alarm.

"What?"

"Now that you know it was just your guys' lack of communication that ruined the relationship, would you want to try again?" It hurt to ask, but the way he sadly looked at the ground after I said they should've talked things out made me think he had some regrets. Sasha was also seemingly miserable without him. Maybe they were meant for each other or something. I didn't know.

Grey grabbed my face in his hand firmly and looked me in the eyes. "Sasha and I tried and failed. It's over. So please, don't ever repeat something as silly as that."

"So you want to stay with me?" I asked.

"Very much so. Do you want to stay with me?"

"Definitely," I said.

"Glad that's been settled then," he said and smirked.

"So you want me?" I teased.

"Oh, you're still going on?" He lifted me as the bench below us turned into a wall. He laughed and said, "This looks so stupid being out in the meadow, but you've given me no choice."

"What do you mean?" I asked. He pushed me against the wall and encased me between him and it. "Okay. Ask me again, Hazel."

I grew shy. "Ask what?"

"You don't remember now?" he said with a small smile. "I'm pretty sure you do." His face inched closer to mine. "Ask me, Hazel."

"Do... do you want me?"

He placed a kiss on my lips, my nose, and my chin and trailed them down to my neck. It caught me off guard, and I tensed up. He gently kissed every inch of my neck. My breath was uneven when he made his way back up to my lips again. I entangled my fingers through his hair as he wrapped his hands around my waist. "Do you believe me now?

"Mhm," I breathed.

"Good," Grey said.

We kissed softly for another second and then leaned our foreheads against each other. His eyes looked back and forth between mine, and I giggled.

"You're so cute," I said.

"You're so pretty," he retorted. I closed my eyes and listened. The animals began playing behind us.

"Can we go play with them?" I asked.

"I'll race you!" He let go of me and began running.

"No fair! My back was to the wall!" I said, chasing after him.

# Chapter 25

Our relationship grew deep as the summer went on. We went on many more dates in the real world without interruption and in the dream world. By the time school started, everything was very different in all the right ways.

The first change was that I told Mindy, Arnold, and Andy that Grey and I were dating. Arnold was the only person who seemed surprised by this or sounded the most surprised through text. His exact words were, "Hold up. Wait. She's with him. He's with her. He made her miserable... yet they're together?!?!?"

Mindy said, "Yep, saw that comin'," and Andy said, "Yeah, me 2."

We had several hangouts with each other to get to know Grey better over the summer. Grey even bought his friends along so we could all get used to one another before school started. And by the time it did, we all sat at the same table at lunch, much to the surprise of everyone else around us.

And Evan and Sasha were allowed nowhere near it.

Unlike Sasha, the popular kids were nicer than expected, and it was Grey's doing. They would've been out of his group if they even tried looking at us sideways.

In September, Ana and Blake had their first anniversary, and Grey threw them a little party at his house.

"Your boyfriend has been redeeming himself," Ana whispered in my ear as he brought out a homemade cake that he and I baked together. "Even Andy has warmed up to him." I looked over at Andy, who smiled as Grey acted like he was about to drop the cake.

"I'm glad," I signed.

"What are you glad for, cute thang over there?" Grey asked as he placed the cake on the marble kitchen counter.

"Stop," I signed shyly. He smirked and then gestured to Ana and Blake. "Come cut your cake," he said.

"Thank you, guys!" Blake said.

"Seriously," Ana said and kissed my cheek. Mindy kicked my leg lightly and whispered, "When will this be, you guys?"

"June," I signed.

Before Mindy could say anything, there was a knock behind us in the kitchen doorway. We all looked, Ana and Blake mid-cut, to see a man standing there. His hair was gray, and his eyes were a bright blue.

"Hey, Dad," Grey said.

"Hey, bud," he responded. "Sorry for interrupting; I just wanted to say hi. Especially because I heard his girlfriend was going to be here." His eyes scoured our faces as he tried to guess who it could be.

"That's Hazel," Grey said, pointing at me. "Hazel, Dad. Dad, Hazel."

"Nice to meet you," he signed with a bright smile.

I know my whole face probably lit up. "You know ASL?"

"My mom went deaf while I was young. I had to learn so that we could communicate with each other. Mainly so that she could yell at me with her hands." We both laughed.

"That's so cool that you learned the language," I signed excitedly. "Is that how Grey knows?"

"Yep," he signed back. "My mom would've killed me if her grandkids couldn't communicate with her."

"Yeah, of course. You did a good job teaching Grey. He knows the language very well."

"Thank you, sweetheart."

Grey then introduced the rest of my friends to him. They all smiled and shyly waved while giving various verbal greetings.

"All right," Grey's dad said. "I'll get out of your hair. I don't want to make Grey feel uncool or anything." He winked at him, and Grey shook his head.

"You don't want any cake?" Ana asked.

"Oh," he said and looked hesitant as he looked at it. "No, it's okay. Thank you, though! See you guys later." He waved, and we all said bye before he walked out of the room.

"Your dad is cool," I signed. "He made my day."

Grey smiled and signed back, "Thank you, babe."

After eating the cake, we swam around in Grey's pool, enjoying the last bit of summer. We played many pool games, which helped us all get closer. Grey and Andy even went off to the side to talk things out.

"Our talk wasn't planned," Grey told me as he drove me home later that night after the party. "We started talking, and I apologized for hitting him, and he surprisingly forgave me. So that was cool." I squeezed his thigh.

"Hey, no," Grey said and playfully slapped my hand away. "I do that." He placed his hand on my lower thigh and squeezed me back. I laughed.

So, that was how the start of our school year went. By the time December rolled around, I'd finally gotten used to my weird but great last year of high school.

***

"I'll meet you in the cafeteria," I signed and gave Grey a quick parting kiss. He walked down the hall with Mindy and Arnold as I walked into the bathroom. I quickly did my business as the noises in the hall grew quieter as everybody headed to the lunchroom. The bathroom door opened and shut right as I zipped up my pants and flushed. I walked out of the stall to be greeted by a stone-faced Sasha. She folded her arms and leaned on the wall perpendicular to the sinks.

"Are you happy?" she asked.

Her words instantly made me feel dread flood through my entire body. Her hair looked frizzy and wet, and her eyes were bloodshot as if she had just finished crying. I knew she wouldn't understand me, so I turned and washed my hands, still looking toward her. She watched me intently and waited until I grabbed a paper towel to dry my hand before speaking again.

"Well, I'm not happy," she stated.

I tried to show on my face that her misery didn't make me feel good. Far from it, actually. Finding out who she was from that one dream alone gave me a soft spot for her. Then again, being in someone's subconscious, especially that of the person you know, was always very personal. I always felt something after leaving people's dreams.

"Do you know why?" she asked. I shook my head. So she answered, "Because you swooped into my life and took everything from me." She pushed herself off the wall and took an angry step toward me. "You took Grey, you took my place in our friend group, you made everyone hate me, and you're getting off scot-free." I tensed up, already knowing where this was going to go. The look in her eyes, the way Sasha clenched her fist at her side, the confidence of her steps... she was about to fight me.

I don't know much of what happened afterward. I didn't feel anything, and I know Sasha got some good hits. But I was just as angry as she was. And I wasn't going down without a fight. My knuckles hurt when it was over, and her face looked like it was beginning to bruise.

"I hate you!" she screamed. She left the bathroom while crying as I stood there in shock. My adrenaline slowly ebbed out of my system. I looked at my face and accessed the damage. With each bruise and cut I saw, which weren't too many, the pain came into existence. I shakily pulled out my phone and texted Grey, "Come bathroom."

When I heard him knock, I opened the door. His eyes widened immediately, and he grabbed my face in his hands. "Who did this to you, Hazel?"

"Sasha. We fought each other."

His expression hardened. He pulled me over to the sink and gently washed my face. Tears fell out of my eyes as he did, followed by hiccuping sobs. He held me close to him as we walked out of the bathroom together. "It's okay, babe. It's okay." We walked through the primarily dead halls towards the dean's office. Grey knocked, and we heard him say, "Come in."

The minute his eyes met my face, they filled with worry. He stood and rushed over to me, examining my face.

"Who did this to her?" he asked.

"Sasha Bernstein," Grey answered. "They got into a fight in the bathroom."

"Let's get her to the nurse's office," he said. Grey wrapped his arm around my shoulder and pulled them close to him as we followed the dean out of his room. I kept my eyes glued to the ground as we walked down the halls. I wanted to sink through the floor and never return.

The fact that this even happened to me didn't feel real—I still felt like I was in the bathroom, washing my hands, and going to lunch. Sasha wasn't there, and my day went by as usual. But instead, I was in front of the nurses' office, and Sasha was already sitting inside.

"Who did you get into a fight with?" the nurse asked her right as she looked up at me. "Oh. Was it her?"

Sasha didn't bother looking up at me. She remained quiet as the nurse began to work on her.

"You two wait outside while I have a word with Sasha," the dean said and shut the door behind him.

***

The dean called our parents after being treated for our cuts and bruises. I wanted to sink even deeper as they walked into the dean's office with us. Sasha's mom and my mom stood behind us, respectively.

"So, why did this happen?" Mr. Garcia asked.

"She and Grey ruined my life," Sasha said numbly. "They bullied me and excluded me from their friend group." Her lies caused his eyes to land on me. I had no words to defend myself and felt alone in the room, even though Grey and my mom were there.

"And what's your say on this?"

"Sign to me, baby, and I'll translate," my mom said firmly.

"Sasha bullied me for months, and Grey has the text messages to prove it. I don't know why she wanted to fight me. After that, things calmed down for a bit, and I've hardly thought about her." My mom translated for me, and Mr. Garcia nodded.

"Can I see the messages, Grey?" he asked. Grey nodded and pulled out his phone. He showed him the text messages from last year. Mr. Garcia took a moment to scan through them before looking back at Sasha. "It looks like you—and even Grey—were the bullies for some time, Sasha. So how is it that they are bullying you?"

"There was a party where they exposed me and had people boo at me."

"And if you recall, Sasha, you did the same thing to Hazel on a worse scale. I didn't tell people to be mean to you; I was clearing up the bad reputation we gave Hazel. You're acting like the victim when Hazel has been and always will be the victim in this situation."

"Hazel isn't the victim; she took you away from me like the slut she—"

"Sasha!" her mom and Mr. Garcia exclaimed.

My mom looked at her mom with a look that could tear her to shreds. Sasha's mom felt it so much that she slapped Sasha hard in the back of the head and said, "Apologize to Hazel and her mom right now."

"Mom—"

"Right now!" she shouted.

Sasha kept her head down as she mumbled, "Sorry."

"I don't know who taught you to behave like that, but it wasn't me," her mom continued. "You should be ashamed of yourself. She can't even speak. You like picking on people who are easy to pick on?"

Sasha said nothing and kept her eyes firmly planted on the floor.

Mr. Garcia shook his head and cleared his throat. "What happened in the bathroom. Who started the fight?"

Sasha said nothing as Mr. Garcia looked at me. I looked away. I know I should've exposed her, but I felt terrible for her remembering her dream. She seemed so broken and alone. Of course, she did it to herself, but I still didn't want to add to what she was feeling.

"Sasha, who started the fight?" Mr. Garcia asked. "Was it you?" She stayed quiet. "Was it Hazel?" She still said nothing. "I think it was you, Sasha. Why did you start the fight with Hazel?"

"Everyone loves her. No one loves me." I closed my eyes and tried not to feel sympathy for her. But hearing those words outside of the dreamworld hurt more than I could've imagined. She felt this way. She didn't feel loved.

Mr. Garcia looked between us, then at her mom.

"I think this is a conversation we should have ourselves," he said. "Hazel, Grey, and Ms. Jones, would you mind waiting outside, please?"

"Sure," my mom said. We walked outside and shut the door right as Sasha burst into tears.

# Chapter 26

Sasha and I got off with a slap on the wrist, and Sasha was to go to the in-school "therapist" until her Dad chose one he thought was best for her.

I don't know why her mom told me that about her father, tbh. I guess she still likes me even though me and Sasha don't get along anymore, Grey texted later that day. I put my phone down on the table as my mom brought a plate of food over to me. She placed it on the table and raised my chin to take another long look at my face.

"You guys got each other good, huh?" she asked.

"I guess we were both super pissed," I signed. My mom sighed and sat down across from me.

"I'm glad you defended yourself. If the dean tried to give you any punishment for that girl's actions, I would've done everything I could to prevent that from happening. Acting like a child and trying to take you down for defending yourself."

"Yeah, I can't believe her."

"When did this all start happening?" she questioned. "Has she always bullied you?"

"No. It started last year."

"I knew something was happening," my mom said. "You were coming home moody all the time, and every time I questioned you, you said nothing was happening! Why?"

"I didn't want to worry you," I signed. "I wanted to handle it myself."

"Hazel, you're never alone," she said thoughtfully. "I'm here for you. I never want you to deal with any of this by yourself. Do you know how quickly I would've flown in that school last year? You wouldn't have had any stress because I would've ended it fast."

I laughed slightly.

"I'm serious, baby."

"I know," I signed. "Sorry I didn't tell you."

"It's okay." We ate for a moment before her eyes brightened. "I see your boyfriend was there for you the whole time."

"Yes," I signed a little shyly.

"So you two are serious about each other?" she asked.

"Yeah," I said.

She pursed her lips in thought. "So we should have him over for dinner, then."

It made me feel queasy and excited all at the same time. "Really?"

"Yeah. I want to get to know more about him. Boys can get dumb after a while. He needs to know your mom loves you very much and will kill him if he breaks your heart in any way."

I knew very confidently that he wouldn't. People probably would think that's a bold statement coming from a high school girl dating a high school boy, but the subconscious was different from our actual world. We felt so close to each other just by sharing the same headspace every night. It was intimate in a sweet way.

But my mom wouldn't understand that because she didn't know I shared dreams with him. And if I told her, she definitely wouldn't believe me.

***

Again, I checked my outfit over and over. I had healed from the healed cuts and bruises on my face from the week prior for the most part. I wore an off-white knit sweater with black velvet leggings and cute fluffy socks, which I'd probably take off in the evening.

The minute I heard the door, I practically ran down the stairs to get it. My mom beat me to it, opening the door to see someone I hadn't expected—Jared. Like a dear in the headlights, I watched him give my mom a soft kiss on the lips before announcing, "I brought wine."

"Our favorite one?" she asked.

"Of course," he said. "Silly." He directed his attention to me as he handed her the wine. "Hazel! How are you doing, kiddo?" He took two brisk steps toward me and gave me a big hug. Feeling many things at once, mainly hurt that my mom would sneak Jared into our plans without telling me, I gave him a soft hug back. He pulled away and smiled. My mom was back in the kitchen, leaving us alone.

"I'm happy we're finally getting to spend time with each other," he said. "Your mom and I have meant to make this happen for a while." I nodded numbly, not even wanting to think about my mom. I was so angry and upset all at the same time.

The doorbell rang again, and I rushed past Jared to get it. Grey smiled as he saw me. "Hi! I stopped at the bakery to bring by cookies." He held a bag at his side, and before he could hand it to me, I threw myself in his arms, burying my head in his chest. When I didn't let go, he whispered, "Are you okay?"

"Hi, I'm Jared," Jared awkwardly said. Grey let go of me to shake his hand. "Oh. Nice to meet you. I'm Grey."

"Nice to meet you, Grey," Jared said behind me. "I'll be in the kitchen if you guys need me!" His footsteps receded, and Grey hobbled us into the house while shutting the door behind him.

"You're upset because Jared's here, huh?" he asked. I spoke about Jared very little during our dates, but those moments were enough for Grey to know how opposed I felt about my mom dating him. I also admitted it was selfish of me, but I couldn't let go of my dad the way she could.

I pulled away from him and nodded.

"I'm here, okay?" he said. "We're gonna have fun. And I'll try to talk a lot, so there are no quiet moments." I nodded appreciatively. I was starting to feel a lot better, and he hadn't even done much to make me feel that way. I felt an intense burst of emotions for him and reached for the top of his coat, roughly bringing him to my lips. They were slightly cold, along with the rest of his face, causing me to smile. I opened my eyes slightly and saw him smiling, too.

I helped him take his coat off and hung it up in the closet before I led him to the kitchen.

"Hello, Grey," my mom said, pulling him into a hug.

"Hi, Ms. Jones," he said. "I brought some cookies for after dinner."

"Thank you!" she said. She grabbed the bag of cookies from Grey's hand and placed it on top of the fridge. "How was your ride over?"

"It was good," he said. "It's nice out even though it's winter."

"Yeah, it's been mild so far," she said. "Hopefully, it stays that way."

"I have a feeling it will," he said. "This has been a pretty great year, and it's gotta end great." I felt my cheeks warm. He was right about that. Even though the year started with a lot of drama, it turned out to be one of the best years I could've asked for.

My mom laughed. "You've got that right."

"How was your drive, Jared?" Grey asked.

"It was nice," he said. "Got to listen to some good music on the way as well."

"What'd you listen to?" Grey asked.

Grey and Jared began to talk about their music taste and likes while I helped my mom set the table and bring out the food. We finally sat down at the table and began to eat.

"Your food is great as always, honey," Jared said. "Thank you!"

"Yes, thank you," Grey agreed. "You have to teach me this rice and gandules recipe. It's delicious."

"Yeah, sure," she said. "You can come over any weekend you want, and I can teach it to you."

"Yes, please!" he said. He looked so excited. "I've been wanting to learn more recipes so that my dad could finally catch a break on the same meals all the time."

"I can teach you more than just that," she said. "I know a whole bunch of different dishes."

Grey and my mom continued chatting about the variety of foods they enjoyed cooking. Jared interjected every once in a while with recipes he had up his sleeve. I tried not to stare at him too much, but everything about him made me angry. He was nothing like my dad. Why would my mom ever want to date someone like him? And it was still way too soon.

"I wanted you to know that I appreciated how you helped Hazel last week," my mom said after their food conversation died down. "I was impressed with your maturity."

"Thank you," Grey said and seemed slightly shy. "It angered me to see her hurt like that." He nudged me and then kissed my forehead, and then he whispered in my ear, "I like you a lot, cutie. I'm sorry I couldn't protect you from her." After that, it was my turn to be shy. I smiled as I looked down at my plate, shoving more food into my mouth as a distraction.

The night continued with us playing charades and video games.

"Mopping!" Grey shouted. I shook my head as I tried to make a more exaggerated shoveling motion, even trying to throw dirt over my head. He continued, "Aggressive mopping."

I laughed and shook my head.

"Uh... lawn mowing!" my mom joined in. I put my head in my hand and shook my head, much to everyone's amusement. Then, still laughing, I started doing the movement again.

"Oh!" Jared said. "Moving dirt! Uh... shoveling?" I nodded.

"Sweet," he said. "Good job, team!" He raised his hand for a high five, and I met it enthusiastically as we switched places. He stood in front of the TV, and I sat beside Grey. Grey wrapped his arms over my shoulder as Jared started his movement.

After a few more games of charades and Mario Kart, it was finally time for Jared to head home.

"Before I leave, do you mind if I talk to you for a moment, Hazel?" he asked. I looked at my mom, who seemed to know this would happen. She gave me an encouraging nod, and I immediately felt sick. I followed him out of the room into the hallway.

He gave me a sad smile before saying, "I feel you don't like me very much. And I wanted you to know that you have all the power in this situation, even if you don't think you do. Your mom told me very seriously that she won't keep dating any guy if it bothers you. So whether you realize it or not, your mom will break things off with me if you disapprove. So you have the upper hand here. And I'm sorry if any of this has made you feel helpless.

"I love your mom very much. And that means loving everything about her, including you. I know I can never be your father, nor do I ever think I can assume that role. But if you ever decide that you like me in the slightest, I will do my best for both of you. But again, the choice is yours. And I understand whatever choice you make. And you don't have to make it right now; no matter what happens, I loved getting to know your mom, and she is very blessed with such a great kid. You bring her a great deal of pride."

His words echoed throughout my brain and twisted a knot in my stomach. Suddenly I felt guilty. I didn't know I had any say in this situation. I thought he and my mom would be married as soon as I left the house and went to college. But to hear that that's not the case made a sudden weight lift off my shoulders and a newer, more severe weight replace them.

Before, I was bitter that my mom would just let him into our life without giving me a say in anything. Now, my mom's happiness is in my hands. I could tell her that I didn't like him and never would, and he'd be out the door and never come back. That made me feel worse. I'd rather be the one miserable over her.

I took a hesitant step toward him and pulled him into a hug. He hugged me back gingerly and let out a relieved sigh.

I could do this, I thought. For my mom.

***

"How're you feeling?" Grey asked as I walked him to his car. Jared was long gone, and my mom was in the kitchen, packing away snacks.

"I'm fine, I guess," I signed. "I don't like that my mom wants to get with him. I don't get what she sees in him compared to my father, but she's happy. So I'm doing this for her."

Grey pulled me into him and kissed me. "You all know he can't replace your father. But if he's trying his best to be a good dude in this situation, I think you're doing the right thing."

"I wish it felt more right," I signed.

"Try looking at it more positively. That might help."

"How is it positive?"

"How is it negative? He's nice to you, he's nice to your mom, and they both seem like they genuinely love each other, you know?"

I nodded and leaned into him. Being in his arms made me feel better. He held me for a long time. Only when the cold seemingly seated into both of our bones did we give each other a parting kiss.

"I'll see you later tonight," he signed.

I smiled. "See you then."

# CHAPTER 27

Later that night, I phased into Grey's dream wearing a night-gown that looked medieval, and a servant was leading me down a dimly lit hallway.

"Your husband is awaiting you in your chambers," she spoke softly. A small smile overtook my face. My husband, huh?

"He requested you come immediately," she continued. "He says you and him have some important things to discuss." I followed her down the hall until we finally arrived at a heavy, wooden door. Guards pushed the door open for her, and she bowed in front of Grey. He turned and looked at us with a smile. His low white v-shirt exposed the muscles underneath. My eyes lingered as he commanded the female, "Leave us. Close the door behind you."

She bustled out as the doors groaned and shut behind us.

I smiled slyly at Grey. "So... you're my husband now."

"Maybe," he said. He put the piece of paper he was holding down and approached me slowly. "How do you feel being my wife?"

I traced my hand across the exposed part of his chest. "Do you expect me to say something profound? All I can think about is skin."

He laughed and then pulled his shirt off.

"Abs," I said.

"You are so goofy," he said while laughing some more.

I hugged him tightly and pressed my ear into his chest. "And you are so warm."

His arms wrapped underneath my butt and lifted me off the ground. I gasped, surprised. He carried me over to the bed and softly dropped me down.

"It's warmer in bed," he said. He hovered over me briefly, lying down next to me and placing a soft kiss on my cheek. I faced him and traced lines across his chest.

"Enjoying yourself?" he asked, amused.

"You have no idea," I said. He pulled me close and kissed me slowly. My heart rate immediately began to speed up. He was about to deepen the kiss when there was a knock on the door.

He pulled away and asked, "Who is it?"

"Sire, you haven't forgotten, have you?" a muffled voice said from the other side of the door.

"Uh..." he looked at me, confused. "Forgotten what?"

"The enemy is approaching the outskirts of the walls!" the voice said urgently.

"Oh, crap!" Grey muttered. "That was the whole point of the dream. You distracted me."

He jumped up and pulled me with him. "Are you ready?"

I took a deep breath to compose myself. "Yeah, sure. Who are we fighting?"

"The rival King. He's come to steal you from me." He smirked. "But we won't be letting that happen."

"I love how I'm the damsel, but I'll also be fighting alongside the battle."

"You think I'd let you sit back and be helpless?" He scoffed. "Are you a joke to yourself?" I laughed.

"I hate you," I teased. Grey smiled as he began getting dressed appropriately for battle. I followed suit until I realized I had to change out of my gown. He zipped up his boots and then looked at me.

"What are you waiting for?" he asked.

"I have to change."

He paused before letting out a heavy breath. "You... okay. I'm going to go with the guards. Meet me at the front gates." He lingered for a second before quickly making his way to the door and pushing it open. I let out the breath I was holding as the doors closed behind him. I quickly got undressed as the voices receded down the hall. As I changed, the quiet around me got intense. I pulled the armor across my body and fastened it tightly.

I grunted as I pushed open the heavy door, only for a man I've never seen before to be standing with his arms folded across his chest to be waiting for me. He looked smug as he looked me up and down.

"The Queen goes out for battle as well?" he said, pushing himself off the wall. He began to circle me. "You marry him instead of me, the one person who would've put you out of harm's way?"

He reaches for my arm, and I cringe away from him. He looks at me amused as he grabs my arm roughly and brings me close to him. "What are you afraid of, love?"

He attempted to reach for me again, which I quickly evaded.

"Don't touch me," I said.

"You will be my Queen, darling," he hissed. "You have no other choice in the matter." I tried to move away, but my legs locked. He grabbed me and hoisted me over his shoulder.

"Put me down!" I shouted.

"Sir, he's downstairs waiting for her," I heard a voice say.

"Then we'll escape through the roof again," he said.

"Grey!" I shouted. "Help me!" My voice echoed down the halls and the stairway.

"Gag her!" he hissed. Immediately the man shoved a gag into my mouth, and he threw a cloth bag over my head. They also tied my hands behind my back and feet together, although I kicked and fought against them.

"That's better," he said. I couldn't see where they were taking me. I was carried upstairs and to the roof, which was hit by warm, humid air. They stopped short as Grey's voice came from before us, saying, "Where do you think you're going with my wife?"

Finally!

"You took this wench from me," the king below me growled. My stomach felt the vibrations of his voice. "I'm only taking back what's rightfully mine."

"Put her down," Grey said, his voice calm. "I don't want to have to hurt you." He removed me from his shoulder and placed me down on the ground. Immediately I felt the sharp end of a sword or dagger placed against my throat.

"If I can't have her, no one can," he growled. I expected to feel the impact and wake up, but instead, the man let out a surprised yelp, and the blade was off my neck. The bag got pulled off my head, and Grey looked down at me with a smile.

"I'm legit your knight in shining armor," he said. I rolled my eyes playfully. He removes the gag from my mouth and cuts open the ropes.

"Where did he go?" I asked. "And all the men with him."

"I just dreamt up a hole in the ground," Grey said, voice thick with laughter. "They're good, though. They're probably finding a way back up here or to the battlefield." He turns and picks up a bow and a quiver full of arrows. "Are you ready to fight?"

"Definitely."

"All righty, then," he said. He whistled, and out of nowhere, two big birds flew down from the sky. They looked like giant hawks and slightly startled me.

"Don't tell me you're afraid of birds," Grey said as they landed. He pet his giant hawk's face.

"I'm afraid when I don't expect them," I said. I pet mine's face as well, and it rubbed into my hand. I went around it and saw a saddle already attached to its back. I hopped on, and we flew up into the sky. Using my bow, I aimed at the people below, fighting against each other. Then, spotting an enemy as we swooped in close, I aimed and fired.

***

We won the battle.

The made-up king was defeated and sent home with his tail between his legs. He looked back at me, longing, but I just nodded at him in the direction he needed to keep walking in. The made-up people on our side began to cheer and then gradually walked back into the city's walls. Our hawks flew up in the sky while we stood on the battlefield with grins.

"That was so much fun," Grey said.

"I think that's the coolest dream we've had together by far," I said.

"Nah, I don't think so," he said. He took my hand and led me back to the gate, our guards following us.

"No? What's the best dream we've had, in your opinion?" I asked.

"I liked the dream where we were in the ocean with all the pretty lights," he said. My heart warmed as the gates closed behind us.

# CHAPTER 28

I went over to Grey's house weekly; we spent a lot of time with his father. Of course, he wasn't there every time I was over, but it was enough to get to know him more. Sometimes he'd watch movies with us; other times, he'd play silly video games like Gang Beasts or Human Fall Flat, and every time we were with him, it was a blast.

"Grey, did you just throw me, your flesh and blood, into the grinder?" he asked.

Grey chuckled. "Yeah, pops, and I'll do it again."

"He threw me into pink goo," I signed as Grey threw his avatar's hands up in victory. "He spares no one for his wins."

"I see that," his father said. "Wanna team up to take him out?"

"Yes!" I signed. "We gotta humble him a bit."

"Oh, you guys are so on," Grey said excitedly.

We were only to get Grey down once or twice over the dozens of games we played with him.

Grey and his father made me feel at home. And since we liked each other and our parents liked our relationship, my mom and Jared came over to meet Grey's dad. We all had dinner together that Grey made for us. My mom, Jared, and Grey's dad hit it off very well. Grey's dad was a taciturn guy but lovable at the same time—I think I was biased since he could sign, and he often signed with me.

Watching Grey and his father interact was fun as well. I got to see a side of Grey that he only tended to show in the dream world, not the real world. They always made me laugh when they were together.

"Dad, Hazel, Ready for Linner?" Grey called from the kitchen.

"Did he just say, Linner?" Grey's dad asked.

I laughed and signed, "I believe he did."

We walked to the dining room where Grey had prepared a traditional rice, beans, and chicken meal.

"Thank you, babe," I signed and kissed him on the cheek.

"Yeah, thanks, son," his dad said and made a fake kissing face.

"Get away from me," Grey joked.

We sat around the food and served ourselves. Grey's cooking was always super good. He heard it from me countless times. His dad praised him for it as well.

"So, how was your guys' day at school?" Grey's dad asked through a mouthful of food.

"It was good," Grey said. "Typical day at school as always."

"Good to hear," he said. "How about you, Hazel?"

"It was great. I finally finished this ceramics project I was working on."

"What was it?" he signed.

"Nothing special. I made a sculpture of my dog, Timmy."

"Can you take it home?"

"Yeah, after Senior Arts Night at the end of the year."

"Perfect, then. I'll be there to see it."

"Her ceramics is spot on," Grey signed. "She does such a good job."

I kicked his leg shyly.

"Don't be humble, Hazel; we're all family here," Grey's dad said. "You're allowed to admit you're awesome without judgment."

"I'm okay," I signed, slightly smiling, looking down at my hands.

"All right, we'll cut you some slack. We're going to burn her ears off her body if we don't." I laughed and shook my head. "Hazel, if you don't mind me asking, we're you born non-verbal?"

I shook my head.

"Oh, what happened that you're non-verbal now?" he asked, and Grey shot his father a warning glance. He picked up on it and added, "That is if you don't mind me asking. You can say no, honey, if it's super personal. I don't have to know."

I took a deep breath and shook my head. "It's okay. As you said, we're practically family at this point. I don't mind you knowing." I mentally prepared myself to tell the story. "There was one day I went to the gas station with my dad. When we returned to the car, this man snatched me up and held me at gunpoint until my dad gave him money. Unfortunately, my dad didn't have money on him. And instead of the guy letting him get it or letting us go, he shot my father in front of me."

His dad was quiet and leaned his mouth into his hands as I continued. "I felt... I felt my dad's blood splatter on my face." I kept it together. I shouldn't have shared that, but that was a big part of why I didn't ever want to talk again. His blood was all over my face, including my lips. It being there sealed my lips and locked my voice out deep inside of me. "The man ran away and left me alone as my father died in front of me. And as he died, my voice figuratively died with him." I was on the brink of tears, but I didn't let them spill. "And that was that." Grey's fingers locked with mine tightly.

I looked at his dad. His eyes filled with tears. "I'm so sorry something awful like that happened to someone as brilliant as you. I'm sure that man can't live with himself, can't even sleep at night due to the pain he caused you."

I nodded, and my bottom lip trembled.

"I'm sorry I asked, sweetheart," he said.

I shook my head and wiped my tears. "It's okay," I signed. "It's tough for me, but it's been like... 13 years now? It still hurts but not as much as when it first happened."

"Of course," he said. "Did anything come of the case?"

"No. I didn't see the man, and from what they saw on the CC video footage, he was masked and wearing sunglasses."

Grey's dad shook his head and sighed deeply. "One day, you'll get justice."

"No offense, but I've believed that lie for too long. The man got away with it. He'll die peacefully while my dad died a morbid death."

The table went quiet for a bit, and I knew I had to break the tension. "I don't want this to be how I leave. Do you guys want to play some video games before I leave after we finish l-i-n-n-e-r?"

"Yeah," Grey said. "I'm down."

"Same," Grey's dad said. "We have to beat Grey up in Gang Beasts." I smiled through my tears, which were slowing down.

"You heard that, Grey? You're on." He smiled, grabbed my hand, and kissed it softly.

***

After we played games with his dad, I broke down when I stepped into Grey's car. It was as if he expected that all along. He held my hand for most of the ride home until we stopped outside my house.

Grey held me in his arms as we stood outside my house for a bit. He kissed me many times as I started crying and steadied me. I didn't know what I'd do without him.

"I'm sorry he asked," Grey said. "I should've told him to drop it."

I pulled away and signed, "It's honestly okay. I didn't have to tell your dad, but I did. I didn't mind that he knew. I like you and your father a lot."

"I'm glad," he said sadly. He kissed my lips and my tear-stained cheeks. "I'm just embarrassed that he made you cry."

"He didn't. The man who killed my dad did."

Grey nodded solemnly. I leaned my head back into his chest. His arms remained wrapped around me tightly until I finally got my bearings. Even after I composed myself, I let him hold me until it became too cold to stay out again.

"Are you sure you're okay?" he asked as I pulled away from him after kissing him.

"Yeah," I signed.

"You don't want me to stay for a bit?" he asked. "I wanna say hi to your mom anyway."

"She's out on a date with Jared, and you know her rules."

"Yeah, no, me without her there," he said and then joked, "I'm a little bitter about that." I playfully went to hit him, but he quickly grabbed my hand and pulled me into another kiss.

"I'll see you soon," he whispered against my lips. I nibbled his in response.

***

I knew I was in the wrong dream. I felt discomfort as I walked down a winding path in the woods. The trees swayed eerily as leaves crunched underfoot, and the sun looked like it was just rising in the sky. I thought the path would never end and that I was stuck in someone's nightmare until I heard the sobs.

They weren't coherent, nor did they come from a tangible source. They just echoed throughout the forest as I continued walking. Every time I thought I was getting close to the person crying, they'd sound further away. I tried to keep my cool as I continued forward. If I freaked out, the person's subconscious would overtake mine and make it harder to control their dream. It's happened in the past when I didn't understand how my dream traveling worked.

If I wanted to get out of a dream, I had to be calm and collected.

The sun was the only thing that guided me as I walked forward. It was still low in the sky, but it was enough to shine through the trees so I could see where I was going. The path I was on twisted slowly in several directions as I walked. I followed it closely, knowing it would lead me to the dreamer. Then I could wake them up and finally get out of his horrid nightmare. The anguish in their sobs resonated deeply in my chest.

I made the plan in my mind. I'd find the dreamer, maybe try to comfort them, and then manage to wake them up out of their nightmare.

Then I could finally meet up with Grey. Thinking of him made my heart lighten. I could do this. I'd see him soon. I wanted to do more fun things with him, like in the last dream and maybe give him a kiss or two or seven. So, with new conviction, I marched forward more confidently.

Once I finally came across freshly overturned dirt at the end of the path, the sobs stopped. I was curious as to what objects were hidden underneath the earth. Maybe it was the thing making this person cry. So I bent down and began digging. The dirt was wet as I combed through it. It wasn't until I finally touched the ski mask and sunglasses that I knew whose dream I was in.

My dad's murderer.

My hands shook as I held the mask and glasses in my hands. Footsteps crunched behind me, and I froze. I knew he was nearby. Panicked breaths filled my chest as I came to understand what I had to do. I wanted to believe Grey's dad, that I'd finally get justice. That meant I'd have to look at the man straight in the eyes. He had no mask. No sunglasses. I would see him. I'd finally know who he was.

And somehow, I would prove to the world who he was and that he killed my dad. Somehow. That I owed my dad.

"If I could take it back," the man started, his voice slightly away from me, "I would."

Anger flared through my body. "But you can't. My dad is dead because of you. And this is what you did, huh? Hid the evidence in the woods."

"I didn't know what else to do."

"How about turning yourself in?" I said.

"It wasn't that simple," he whispered. I couldn't believe him. Wasn't it that simple? It was. If you couldn't handle the consequences, you shouldn't have done the action. But I wouldn't bother to waste my breath telling him that. He was a pathetic man who I wouldn't let get away with my dad's death any longer.

I stood. I prepared to turn around, grasping the mask and glasses in my hands.

"You don't want to do that," he said. "You don't want to see me."

"I won't regret seeing who you are, finally. But, you! You'll regret finally facing the girl whose father you murdered."

"Hazel, I already know what you look like... please, don't..."

I stiffened. I slowly began to turn around as I asked, "How do you know my name?" I saw his silhouette in the trees, the bright sunlight obscuring his face. "Who are you?" My voice was rising. "I asked, how do you know my name?! Tell me how!" The woods went still, and he didn't move. We watched each other in complete stillness. He finally relented with a heavy sigh, and his head fell in his hands. I couldn't move as I waited for him to do anything. My heart was pounding in my chest so hard that it hurt. I wanted to scream at him to get it over with, but I couldn't bear to say anything else. Finally, he removed his head from his hands and straightened his back.

He took two slow steps into the sunlight, and my heart shattered into a million pieces.

"No," I sobbed, stumbling away from him.

He dropped the accented voice and said, "I know your name because you're dating my son."

THE END

~To be continued~

www.ingramcontent.com/pod-product-compliance
Lightning Source LLC
LaVergne TN
LVHW021509140926
848035LV00047B/1012
* 9 7 8 2 7 0 7 1 2 5 3 0 9 *